Reduced Load Ballistics Research Manual

Research Guide and Study Results for Novel, Silent, Low Recoil and Gallery Loads

The Lost Art of Reduced Sound and Recoil in Sub and MOA Reloading, including Military, Law Enforcement, CAS and many other research applications

Tom Halstead, AAGSR

www.ReducedLoads.com

American Association of Gunsmiths and Reloaders

www.AAGSR.org

INFINITY
PUBLISHING

AAGSR.org

Reduced Load Ballistics Research Manual: Research Guide and Study Results for Novel, Silent, Low Recoil and Gallery Loads
1. Ammunition Reloading.
2. Ballistics Research

ISBN 978-0-7414-6516-0
Library of Congress Control Number: 2011923404

Printed in the United States of America

Published March 2013

INFINITY PUBLISHING
1094 New DeHaven Street, Suite 100
West Conshohocken, PA 19428-2713
Toll-free (877) BUY BOOK
Local Phone (610) 941-9999
Fax (610) 941-9959
Info@buybooksontheweb.com
www.buybooksontheweb.com

Warning and Disclaimer

All loads, including reduced loads, are potentially dangerous, and none of the loads studied in this manual have been tested for safety by the author, AAGSR, Smith Tactical, or the publisher. The purpose of this manual is to report on standard deviation results, pressure traces and techniques from tests conducted at Smith Tactical, and commissioned by AAGSR, of commonly used and openly reported and referenced, via widespread primary manufacturing sources, reduced, or "gallery" loads.

The correlation between standard deviation and "safety" is unknown and speculative. "Street" logic holds that lower SD's (more uniform velocities) mean tighter groups, which means more consistency, which means greater safety. That chain has not been proven in enough research studies with enough trials to even give goodness of fit of the data. It is known that most reduced load accidents happen not because of formula, fill or detonation, but because of double charges, loading errors, or other extreme overloads.

The publisher and author specifically disclaim and warn against using any information in this book unless you are a researcher with a ballistics lab where an unexpected explosion can only take place in an explosion proof chamber (EC). Use this information at your own risk, and do not attempt loading or reloading unless you have taken the recipe from a published manufacturer's load table, NOT this or any other book or source reporting experimental pressure results from a controlled, explosion proof lab. **The reader agrees with and acknowledges these warnings, and if such reader cannot resist attempting novel loads, will only do so using a tested, certified explosion proof chamber or test facility, and never in the field, or holding the test gun, and holds this publication harmless.**

There is a great need for new researchers in this field, and many career opportunities, however this is NOT a career to start in your yard or garage without an EC, and requires a safety controlled environment, regardless of location. The purpose of this book is to report on research results and encourage research, NOT suggest, certify or encourage the use of any load.

This book is a research report, not a recipe book!

Throughout this book, all research loads have been EC tested-- EC stands for Explosion Chamber, and no ballistics lab, including home labs, is complete without a certified, approved EC, compliant with all local zoning and safety regs.

DEDICATED TO ALL SUPPORTERS OF THE SECOND AMENDMENT

There is a wide and growing divide between the attitudes of Country and City dwellers toward guns and ammunition. In the City, guns are often seen as frightening weapons of violence. In the Country, guns are often seen as tools for fun family events and outings, protection of pets and livestock, bonding between parents and children, survival and hunting, target practice, and a symbol of self sufficiency.

Prejudice literally means "pre-judging" something or someone. In other words, judging something without even trying it, or someone without knowing them. In nearly every case where a "gun banner" actually gets out and tries shooting, attitudes change! *People also are only shot like sheep, by armed assailants, if they are unarmed and untrained themselves.*

100 years ago, nearly everyone loaded their own ammo, and often families only survived because of stealth and accuracy. Along the way, recreational gallery loading (also known as reduced loads, whisper loads, youth loads, etc.) was born. With the advent of modern manufacturing, new ammo and reloading/ ammo bans, we've become more and more dependent on long supply chains and complex technology, when we can get "unmarked" ammo at all. With those shifts, much of the old art of gallery loading has been lost.

This volume is dedicated to those research professionals willing to try something new, both for fun, and potentially as a career and business-- developing loads for law enforcement, military, hunting, training, youth, and just plain fun. With fun and research comes a next generation supportive of our Second Amendment Rights, and contributing to our country's LE and Military needs. The dual purposes of silent, low recoil loads have always been to introduce youth and women to low recoil, quiet, high accuracy loads; and at the same time, expand the reach of load development with new techniques and science applied to an ages old art-- gallery, or reduced charge loading.

TABLE OF CONTENTS

INTRODUCTION-- WHAT IS GALLERY LOADING?

Welcome to the fun, exciting and potentially profitable world of gallery / reduced loading research! **What happens when an old lost treasure**, or an old lost art, is rediscovered? For one, a whole bunch of smart folks with good timing find new fortunes, careers and applications for the old ideas. These are called "early adopters" by the Marketing types!

"Gallery Loading" is the old name for reduced charge and primer only rounds that were used in carnivals, street fairs and home/mobile "shooting galleries," where youth, customers and trainees could test their skills at breaking balloons or knocking over tin targets; often with rubber, cast lead ball, or wax bullets. Today's technology has EXPONENTIALLY expanded those applications! **Will YOU have the next new breakthrough idea in reduced loading research?**

We've expanded the definition of **Gallery Loading** in this manual to keep up with those new applications. Today, "gallery" applications, in the broadest sense of the term (adding high accuracy and special applications to reduced loads and/or unique projectiles) can range from expensive Simunition ™ rounds, to quick draw competition with aluminum barrels, cowboy action shooting and loads, covert assassination and special ops rounds, specialized sniper ammunition, and many other varieties of low recoil, low sound, or special projectile ammunition.

The market for these products and services is skyrocketing, with demand far outstripping the meager supply-- even holding constant the problems of ammunition created by gun banners and anti gun politicians. Gallery loaders from 100 years ago would be shocked to see Thompson Center, Magnum Research, Taurus, Comanche,

Bond and many other manufacturer's handguns that use rifle or shotgun cartridges. Not that the venerable 45 Long Colt is anything to dismiss in any era, but Browning himself would be astonished at 444 Marlin, 30-06, 30-30, .308, .375 H&H, .416 Rigby and 50 BMG "handguns," regardless of load!

That's certainly not to say that the old techniques don't have a lot to tell us. In fact, this manual wouldn't be possible if many of the old documents and techniques had not been archived, and their art kept alive over many decades by **cast bullet loaders**.

When a modern gallery load researcher "discovers" that his little #2 lead cast ball .375 H and H behind 11 grains of Green Dot is far more accurate than his 240 grain cast "bullet shape" projectile, which tends to keyhole with reduced loads (vs. single hole groups at 50 feet for the ball at 1,100 FPS), he's "rediscovering" the introduction of the Minie-- which for many years couldn't come close to the accuracy of simple round ball ammo! (Not to be confused with "ball" ammo-- which is NOT round, and not a ball, but is today a military term for full metal jacket ammo that is not incendiary or metal piercing; or "ball" powder-- a shape of the powder itself).

Technically, the Ballistic Coefficient of a round ball compared to a sleek spire is gosh awful-- and we'd expect terrible looking, drag induced ballistic curves at distances. But bear hunters know that the sleek spire point looks sexy at 300 yards, but has terrible terminal ballistics, low retained weight, and is easily deflected by brush, compared to the "bad" BC round nosers in 375 H&H and 416 Rigby used on some of the most dangerous game on the planet.

In the above example, our gallery loader used **0000 buckshot (.380) swaged to a Cerrosafe bore-gauged .377 (#2 Lead), in a CZ bolt 375 H and H, with a light crimp**

in the center of the ball. Who says a 375 H and H won't work on a Skunk or Coyote? Tired of seeing all your powder blow out of the barrel of your .308 T-C in an impressive display of flame throwing, that barely misses your forehead? Give the above a try in its .308 version!

But won't 11 grains of Green Dot in that big a cartridge risk explosive detonation? Shouldn't he have used 17 grains of Blue Dot instead, with Pufflon, or a 1/4" wad of Dacron batting? Or 10 grains of Red Dot, which burns faster? What about Unique, Herco, or SR-4756? Isn't SR 4759 specifically intended for reduced loads?

Don't get discouraged! This manual will answer all those questions, and many more, **until you've become the top Gallery Loading research expert** in your State, Region or Country-- the "go to" person for hundreds of specialty applications, **including sniper/ dangerous game-level accuracy**! Safety can't be guaranteed with Standard Deviations, but at least caring about them can make any researcher or loader more cautious.

Objectives and applications always drive load development in the real world, but there also are scientific/ research questions we'll try to answer, or at least ask, in the advanced topics sections. If your interest is a step up-- as in Engineering, Ballistics, Risk Analysis, Modeling, advanced Load Development, etc., you'll love these sections. You don't have to be an advanced techie to get a lot from this manual, but we've included enough science to please you if you are, even though basic researchers also will learn a lot-- we certainly did. In fact, there is no area as **ripe for research and innovation, and potentially great profit and new patents**, as reduced loading today.

In addition to LE, Military, Snipers, Covert Ops, rifle loads for pistols, youth, cowboy loading, shotgun technology applied to rifles and pistols, subsonic shotgun loads, etc., there is enough pure ballistic science, chemistry and pure research in gallery loading to bring something for everyone, at every level of loading. Our largest audience for the pre release e-versions of this manual have been experienced research reloaders and small, including home, ballistics labs (not forgetting that explosion chamber!), who found a niche for reduced recoil loads, services, tools, packages and equipment for fun, plinking, testing guns, special contracts, varmints, competitions, and just plain less expensive, lower recoil, yet still impressively accurate rounds. Many have started **exciting new businesses and careers in this exponentially growing field, regardless of slumps in other areas of the economy**.

The reader who put 6 .357 magnum wax rounds into the chest and back of the homeless burglar who broke into his home armed with a screwdriver, has another view of gallery loading-- staying out of jail. We consider his philosophy a prep-step to peppering and tasering the critter. Not as bloody, but very effective, without having to hire a lawyer. Also works well on the neighbor's dog or cat hanging out in your yard, if you're too nice a person to whack the bugger. Other folks pop Javalenas and other pests that don't taste good enough to shoot with "real" ammo. And **for city slickers-- the "pop" is about as loud as a hand clap**-- nothing for the neighbors to fret over.

A properly constructed 12 gauge "Urban" load can put an ounce of **FF buckshot behind 10 grains of Red Dot or Titewad, and with proper use of wads, felt and over-shot cards**, load a home defense hull that sounds more like an air hiss than a bang-- at between 850 and 900 FPS, well below the 1130 of the speed of sound, but at all normal self and home defense ranges, equally effective and deadly!

With the population aging, and 65 million boomers retiring, and ammo at record prices vs. a worthless dollar; accurate/ reduced recoil/ inexpensive alternatives to factory ammo is an idea whose time is NOW, **especially if you're looking to launch or expand a home lab based Reloading, Ballistics, Services, Tools, Equipment, Package, or Gunsmithing business (see the AAGSR pitch on the last page of this book for credentialing info for your lab)**.

As partners with the **largest specialty load development/ Gunsmithing contractor and reduced load ballistics lab in the U.S.** (www.SmithTactical.com), we found that there was a lot of confusion in the literature and on the web surrounding reduced loads, *a lot of dangerous misinformation, and very few tested resources.* This manual is the result of **decades of practical experience in reduced load development**, and a lot of research, and will hopefully fill that gap, while presenting fresh new ideas and questions **for your own successful business or lab**.

From techniques to tips to load research, this guide is the ***most complete resource available anywhere on Gallery and Reduced Loading.*** As a bonus, we've included sample advanced sections on sniper, LE, Covert, and Military loading (Smith's major lines of business), for those of you who'd like to **evolve a fun hobby into a profitable new business or career, serving our military and LE with testing lab services.**

Sniper, or match loads, are simply highly specialized cartridges designed for **extreme accuracy in a specific firearm, for a specific application**. Since these are normally associated with maximum loads to achieve stabilization and the velocity and kinetic energy "sweet spots" of the ballistic curve, our approach in designing highly accurate, energetic, **reduced** loads is unique in the industry. As just one example, our proprietary, patent pending VentriLoad™ goes beyond silencers and suppressors, to create a **distracting sound signature away from the sniper team**, attracting enemy attention to the wrong place, and exposing additional enemy combatant, sniper or terrorist positions, as just one example of many.

Whether you're just starting out in loading, or are an advanced Engineer, there is something here for you. Also visit us at www.ReducedLoads.com or email us at Mail@ReducedLoads.com if you have questions, suggestions, or ideas, and we'll help you stay up to date for a lifetime!

<u>FOREWORD: HOW TO USE THIS GUIDE</u>

Most of this manual is written for the intelligent, research oriented reloader or gunsmith, with a home or professional ballistics lab. See the photos in chapter 13 for a typical "basement" ballistics and load development lab. The engineering and science chapters extend the subject to Engineers, Hunters, Snipers, Armorers, Advanced Gunsmiths, Teachers, Manufacturing Researchers and advanced load developers, who'd like to get a little deeper into the science, chemistry, math or ballistics. **We've also added enough sweeteners to bring the basic loader up to quite advanced levels about how standard deviations and other tests work in general.** We'll keep you up to date with frequent additions on the accompanying web sites.

In either case, read the safety info in Chapter 2 and the basics in Chapter 3, before trying any of the loads in the load sections, even in a lab!

We test all our loads in our ballistic labs extensively, and have not included ANY experimental load formulas that don't correspond to an already **published and tested** (even if out of print) load from a bullet or powder manufacturer. Some of the ideas here might seem quite foreign, unique or new, but it is only because they come from a less known specialty such as old gallery documentation and studies, SOCOM, LE, simunition tests, casting, cowboy shooting, quick draw competition, cast bullet tables, international sources, etc.

If you're planning on starting a new research oriented reloading, ammo design or manufacturing, or gunsmithing business, <u>read this manual from cover to cover!</u> It has ***many hints and tips specifically intended***

for those looking for a new business or career in this exciting new area of loading and load development.

A unique aspect of this book is our own extensive database of Standard Deviations-- never published before, and not available anywhere else. These correspond to our own 100+ round chrony and fully instrumented pressure curve tests (also listed in the load tables) of all previously published reduced loads. This will help your own evaluation of both safety and accuracy before selecting loads for further testing.

Chapter 1. Applications-- Gallery and Sniper Loading

What we are calling Gallery Loads in this manual includes many other traditional and newer synonymous designations and terms, such as:

-- Youth Loads

-- Whisper Loads

-- Reduced or Mouse Loads

-- Low Recoil Loads

-- T-C Loads

-- Cast Loads

-- Assassin's Loads

-- Silent Loads

-- Practice Loads

-- Less than Lethal Loads

-- "City" or Urban Loads

-- Sub Sonic Loads

-- Explosive and Pyrolant Loads with reduced initial stage propellant (see Ch. 12)

--"Squib" loads (actually, a squib is a bullet that gets stuck in your barrel due to a reduced load, and is a very bad thing, but many years ago ultra light loading also was termed squib loading ! One reason reduced loads are rarely/ never used with jacketed bullets is that it is one of life's true nightmares to try to remove a jacketed squib from a barrel. Lead doesn't present quite the same level of nightmare for removal, **but can still kill you if you fire a second round into the first**).

... as well as many terms derived from the **applications** of reduced/ gallery loading. Some of these **applications** include:

-- Law Enforcement Training

-- Self Defense, personal protection, home defense, less lethal

-- Animal and Crowd Control

-- Low cost practice rounds

-- Simulation and Modeling

-- Research, load development, modeling and testing

-- Recreation, plinking and fun

-- Pest Control

-- Varmint and small animal hunting

-- Cowboy action shooting and competition

-- Quick draw practice and competition

-- Range shooting with lighter bullet traps

-- Covert Operations and special ops/ assassination

-- Rifle and shotgun loads in pistols and vice versa

-- Black powder and reenactment hobbies

-- Low sound signature sniping

-- Lower recoil practice

-- Reduction of lead and powder fouling with cast bullets i.e.: "magnums" -- generally defined as loads capable of delivering 2500 FPS consistently over all bullet weights-- requiring gas checks. (Many lead/antimony cast bullet formulas draw the gas check line at 1,600 FPS or lower, depending on hardness-- see the manufacturer's specs). This manual deals with the other end of the spectrum-- mostly subsonic or close to subsonic loads between 500 and 1450 FPS (1132 or under for subsonic). We work with many levels of reduced loads, and of course they also depend on the size of the cartridge and type of firearm. A 4,000 FPS projectile will be "reduced" at 2,000, whereas many consider 1130 an important cut off (speed of sound at 72 degrees). It really depends on objectives.

It is important that you learn the terms and applications, as well as categories of gallery loading, especially if you are planning on making a little extra money, or creating a new business, loading, or selling services, products, tools, equipment or packages in this area. **These many applications will become your market niche opportunities, customers and business base!** In addition to these applications, gallery loads are also categorized by bullet type, pressure, load, case, powder/ propellant type and other technical details. Some of the **projectiles** used in gallery loads include:

-- Traditional bullets

-- Cast bullets

-- Bullets made of wood, plastic, wax, rubber and other materials

-- Loads that shoot only coarse grain gunpowder as a projectile to break balloons (in which case **the powder is a**

projectile, NOT a propellant-- be CAREFUL of this distinction to avoid overloading!)

-- Multi-stage or component projectiles (including reduced loads that have a very quiet/ low recoil initial stage, with pyrolant propellant next, and possibly an exploding tip, see Ch. 13 for examples from Smith Tactical/ AAGSR).

-- Loads that shoot non-lethal objects and substances

-- Loads that shoot or contain poisons, toxins, and other deadly projectiles

-- Round or minie balls, often cast by the shooter

-- Pyrotechnic, simunition, paint and other specialty projectiles

-- Projectiles that contain expanding gases, explosives, additional propellants, poisons, and other exotic military and covert applications

These loads use both powder and primer-only propulsion (including all categories of rifle, pistol and magnum **primers**, as well as 209/686 (European) and 209A **shotgun primers**. Primer-only shells usually have either enlarged pistol/rifle primer pockets drilled out 1/8" (to reduce primer back out when there is no powder to push the primer back flush during firing), or larger 209-sized pockets drilled out (from .38 and .45, to larger rifle shells). (Note that 1/8, or .125, is still slightly tighter than the standard US .175 or .210 primer diameters). The 209 system is used a lot in quick draw shooting practice with wax bullets (see Chapter 13), to avoid shooting one's foot off. One can buy the shells already drilled, **or drill the pocket out to a 209 calipered dimension of between .243 and .244, or just under 1/4" (.250). 15/64 is .234; 6mm is .236**. For practical purposes, the commercial guys use a slightly honed down 1/4" bit, (1/4 is a wee bit big, but ok) and the primer "falls out" pretty easily (it's not an overly snug fit, purposely). The commercial guys also chamfer the flash hole to create a nice smooth lip after drilling.

Gallery/Reduced loads **not only use rifle cartridges in pistols, they also extend shotgun technology to pistols and rifles**, and conversely create "bullets" and other reduced loads for shotguns, **particularly .410/45LC**. 12 gauge "recoil reduction" is often just a few less pellets-- .410 also now has an amazing array of new options. Of course 12 and 20 also have many new expanding and bullet like projectiles that make "slug" guns-- previously a joke at longer ranges-- truly 300 yard capable. **12 gauge "home defense" loads for city dwellers include subsonics based on powder and wad rather than just reducing pellet number.** The combination is important to create a "hand clap" silent load, since **209 primers all by themselves can be quite noisy, depending on technique**.

Because we're driving an ounce of shot at under 1,000 FPS, we recommend Bismuth, but we've gotten good Standard Deviations (tight loads-- see below) with 11 grains of Titewad under an ounce of lead and a chrony of under 900 FPS! **We use a minimum of F to FF shot-- blasting 24 to 28 .22 caliber size pellets at close range, with very low recoil and report.** BPI (see resources chapter) publishes a number of reduced 12 loads, and is a fine source of the exact wads, cups, overshot cards, etc. you will need for a very stealthy load. Remember, these are 50 FOOT loads, NOT 50 yards!! Reducing your pellet count for close range means very deadly up front, but is NOT intended for a cloud of shot downrange. The tradeoff is immediate mass vs. dispersion, **and you need to drive immediate mass with reduced loads**.

209 primers in a bored out pistol cartridge pop in and out with just finger pressure, and don't require decapping or primer loading with a press or hand priming tool. The

primer is usually inserted first so that inserting doesn't just push the projectile back out when there is no powder in the shell. Overdrilling the pocket is important not just to avoid primer back out as an irritation (revolvers will get stuck cylinders with rifle or pistol primers, and sometimes need gunsmithing to remove the shell), but non-flush primers can also be dangerous in the pistol and rifle primer categories.

For example, an incorrectly seated primer can cause a semi to fire out of battery, and blow the shooter's hand up, or worse. **Of course reduced loads are not intended for semis in general**, as they normally won't cycle the action due to load. Most gallery loads are for revolvers, single shots, and bolt actions. If you're loading ANYTHING for semis, double check that primers are flush! Notice we called the reduced load primer "back out" an irritation rather than a danger, because the gallery primer popped out AFTER firing, whereas an improperly seated primer in a semi is still "live."

Also remember, **revolvers are slightly more pressure forgiving than guns with closed actions**, such as the Thompson Center, because some of the gas (a lot actually) escapes out the front of the cylinder, as well as the muzzle. The guys who have blown their thumbs off-- a good number of them-- holding the Smith 460 incorrectly, know all about cylinder blast, the hard way!

Speaking of pressure-- **DO NOT try reduced loads in antique, damascus or other non modern weapons, ever, even in your research lab's Explosion Chamber (EC)!** With known pressure tolerances between 10,000 and 50,000+ PSI, modern loads are formulated for, and matched to, modern guns. Also, NEVER reuse a reduced load cartridge to load a full (even starting) load again-- you've fire formed the reduced load case to a different headspace and chamber sequence.

Some patient souls do use reduced loads in their 1911's, but need to load each round individually when the action won't cycle due to too low pressure (but don't forget the primer/out of battery issue). Same goes for LE, or animal and crowd control with rubber, reduced recoil lethal LE, and less than lethal shotgun rounds-- they are normally used in a double barrel, pump or single shot rather than a semi shotgun for the same reason.

Once we get beyond the primer-only applications and categories, we enter the **vast field of black and smokeless powders used in reduced/ gallery loads.** We'll cover black powder/ pyrodex a little later, as the majority of gallery loads today use some version of a smokeless powder -- either a fast burning shotgun or pistol powder, or a specifically designed faster burning or more stable velocity rifle powder. Like anything else, there are trade-offs, risks and rewards, depending on experience and care. The next chapter covers some of both myths and real dangers of reduced loads, as does the advanced ballistic/ engineering chapter.

Without getting into specific loads yet, here are **some of the most popular SP's** for gallery loading (generally under 25 grains, sometimes 6 to15 or less):

-- Red Dot, Green Dot, Blue Dot, Titewad

-- Unique

-- 700X

-- Herco

-- PB

-- SR 4756

-- 630, 748

-- SR 7625

-- Bullseye

-- 231

-- HS-5

-- AL-5, 7, 8

-- SR 4759

-- HP38

-- 2400

-- IMR 4227

-- RX 7

-- IMR 4198

-- IMR 4320, 4350, 4064, 3031, 4198

-- H 4895

This list is hardly exhaustive (more later in the Load Chapter), but gives a good feel for the large range of SP's (Smokeless Powders) being used in reduced/gallery loads today! We'll comment much more extensively on this list below.

Another way to look at Gallery Loading categorization is of course by caliber and firearm. This naturally intersects with all the other categories. For example, you'll find a lot of .45 Long Colt and .38/ .357 being used in the cowboy and quick draw fields, for obvious reasons. Less obvious, due to conservation of funds and simplicity, **many cowboy shooters like to "match" the caliber in their revolver, derringer and rifle**-- often a lever action .357 or .45 Long Colt.

On the other hand, if you sort by the gun itself, the lever action leaves strictly .357 and .45LC to get into a bunch of other popular applications, from 30-30 for deer, to 444 Marlin and the reliable old 45-70 low pressure spec "Buffalo" gun. When you leave the round noses and get into spires, you can have a **"chain" detonation problem of a tip igniting a primer in a lever action rifle tube, hence the traditional popularity of the rounder 30-30**, and the recent "Lever Revolution™" (Hornady) type rounds, with softer tips.

This tends to limit lever calibers (at least in mass popularity, not wildcatters!) to frequent popular loads like the 30-30 or 45-70, in addition to cowboy. Of course there also are 30-30 Thompson Center pistols, which is a whole additional field of gallery loads for reduced recoil in a very popular caliber. The Thunder 5 revolver has an optional 45-70 cylinder, and you can also get a Magnum Research revolver in 45-70, as well as many single shots. Depending on how hot the load, these, and the 444, **can be a handful, making them worthy candidates for reduced loads**. Shotgun loads are another entire category, and can be both very deadly, yet very quiet, with very low recoil.

These calibers tend to "group" themselves naturally, as many loads are necked down or beefed up versions of others. The 30-06 has been called the "mother" of many cartridge designs, because numerous shells up to the awesome 50BMG use its design elements. A lot of the 45 field centers around the old 45 Long Colt, and even the ultra modern Smith 460 Magnum will shoot both 45 Long Colt and 454 Casull without a cylinder or barrel change.

This is another whole type of "reduced" load--**complimentary calibers**. Add to the fact that you can load the 45 Long Colt as cool as ice or as hot as a Casull, and you have a wonderful range indeed, in both senses of the word. Of course, **most of us will drop a lightly loaded .38**

special into our .357 magnum revolver for a sensitive spouse or child-- another simple example of reducing the load. BUT we **have to be careful using .38 reduced loads in a .357 case if we loaded them ourselves,** for the same reasons we have to watch low powder volumes in larger rifle cases, covered below.

Of course, we can't leave out the many interesting .410 shotgun loads when discussing 45 Long Colt, 454 Casull, 460 Smith Magnum and even 444 Marlin-- there are even reduced .410 shotgun loads using a 444 Marlin fire formed case. (Fire formed meaning you form it by firing it, not by sticking it in a forge/fire, for you aspiring bladesmiths and blacksmiths).

Hunting and sniping center around the size and attitude of the beast, and the thickness of its skin. Excluding 22 mag (rimfire is out of the range of our reduced loading discussions, except for subsonics, primers only, terminal ballistics, .22 nailgun primer loads, and custom bullets); the .223 and 7.62 x 39 can be loaded to be 600+ yard accurate on groundhogs and other "pests". LE and military sniper teams prefer the .308, 7mm, 300 Win Mag, 378 Lapua, Cheytac .408, or 50 BMG, depending on range and type of target (human or vehicular).

Hunters of dangerous game prefer the venerable (and perhaps most versatile) 375 H&H (father of many of our magnum cartridges), 416 Rigby (father of the 338 Lapua), and other big 450-460 and 500 magnums (505 being the starting point for the Cheytac)-- ideal for reduced loads, but needing special thought and care due to cartridge size and headspace requirements (ie-- case expansion forward of the web, requiring thicker designs at that point). *Their normal powders, like 8700, are NOT safe in most reduced loads, even with Dacron. There are a couple 8700 loads that are slightly reduced, even with some decent SD's and pressure traces, but we hate to list them, since they are so dangerous in other applications, and DO NOT transfer well to other reduced loads. We realize there are manufacturers publishing loads like this, but our SD's question the wisdom of their recommendations. Be safe, if you have to try it, do so in your EC!*

"Reducing" a load can also mean pushing a **lower grain bullet** as much as lower grain charge (of powder). Remember though, as the **bullet grain weight goes up, the charge grains generally go down, and vice versa! Also remember the need for more mass with shorter range designs.** In the 375 H&H, one might use a 220 grain for large deer, a 270 for elk and small bears, and a 300 for water buffalo and large bears/ dangerous cats. In that example, the **220 might take 67 grains of Reloader 15, whereas the 270 could take 63 grains and the 300 57 grains as starting (not reduced) loads, depending on the bullet**. Remember: most "reduction percents" given by manufacturers <u>are a percent below MAXIMUM, not starting loads.</u>

Handgun hunters and shooters using rifle rounds are one of the most actively interested groups in reduced loads, for obvious reasons: recoil can be punishing, and a lot of that powder is blown out as flame in a short barrel anyway. **Bullet casters know that even whitetail can easily fall to a reduced load with a good cast bullet, although being humane is of paramount importance when you get beyond rabbits, skunks and squirrels.** "Reducing" in this sense doesn't have to be to 6 grains!

On the other hand, you can load a **.308 handgun with a 100 grain lead ball or cast spitzer with 10 grains of Red Dot, 11 of Green, 12 of SR 7625 or 14 of Herco, and still get 1,600 FPS/ 24K CUP, without breaking your wrist!** These are serious speeds and pressures, without the pain you'd feel trying to shoot a 190 grain .308 Win BTSP behind 40 grains of Varget out of a handgun! Subsonic? No.

Awesome? Yes! (BTW: these are published loads we've tested for SDs, which average around 12, with no pressure curve problems).

Assassin or sopmod loads can be **as quiet as the clap of a hand**, much quieter than the average air gun, and often shoot toxin coated (or containing) round lead balls that leave no ballistics behind. There is a known "noise barrier" with many fast powders that happens between 3 and 6 grains, which relates to powder combustion sound rather than sub or supersonic projectile muzzle velocity. Of course there also are a variety of stun, paint and sleep rounds for animal or crowd control, and even a "skunkball" round that leaves the target "marked" and blind (and easy to identify, if not pleasant), without permanent damage. Even with NO powder, a 209 primer behind a wax or lead bullet, sounds almost as loud as a .22 LR WITH powder. Ear protection level loud? No. Whisper quiet? No.

There is hardly a gun/caliber combination that can't be the subject of creative research for reduced loading, with the exception of already low recoil rimfires. Even there, primer only and other "subsonic" loads are available for sound suppression rather than recoil control, but of course, they also won't cycle a semi.

On the topic of sound, 1,125 to 1,130 feet per second is an important velocity limit for some gallery loaders-- the speed of sound at around 70 F. Even if the reduced load gives a very quiet sound signature via the powder, if the objective is heavily weighted to sound suppression, supersonic projectiles will pose a secondary source of noise above that level. Pellet and air gun shooters know all about this when their projectile is very fast-- some of these hot air guns are louder than a .22. Use your ears, and your chrony, to load to the objective, BOTH in MV and CUP, but not forgetting SD spread and pressure curve overlaps either. Some folks will counter that their objective is more cost control and recoil abatement than noise, but don't forget that sound itself is an important factor in perceived (or "felt") recoil.

The illustrations show the decibel meter we use to chart sound in our research tests of SD's and pressure curves. A good decibel meter is not standard equipment for most ballistics labs, but is a must for the reduced load lab.

For backyard or basement galleries, especially in the city, sound suppression is a key objective (but be aware of lead poisoning and ventilation). In many cities, it is a crime to discharge a firearm within city limits, regardless of the ammo, so check the regs and your specific zoning. With good loading, even moderate music will drown out many gallery loads. But even so, kids talk, and if word gets around that you're running a neighborhood gallery, expect a visit from police, especially in cities run by anti-2nd Amendment politicians. Even some ranges won't permit the use of reduced loads if they have bans on rifle ammo due to backstop integrity or insurance, and even though many of those "rifle" loads are far tamer than a 44 mag pistol round, let alone the 460 or Casull. ***Business reminder: the same ranges that frown on you shooting your reduced loads indoors, will often love to carry them, or special tools and equipment, for sale! See Chapter 13 for pictures of some of the more exotic gallery loads and recipes. Remember, you can't make or sell ANY ammo commercially without a manufacturing FFL! Even if you only do research, it is smart to get your manufacturing FFL to stay on the right side of BATF. Check with your attorney on the most recent regs if you're only setting up a developmental ballistics lab for contract testing, and don't intend to sell or make the rounds you develop commercially.***

We've given this brief range of motivations and applications for gallery loading to introduce the **tremendous variety of**

opportunities in this field. Here are a few quotes from our most active gallery loading readers:

"I gallery load to make a good living in the very specific area of law enforcement training."

"I use reduced loads for my own practice, and to defeat the high cost, noise and recoil of factory ammo."

"I gallery load for self defense and varmint control."

"I'm primarily interested in youth shooting."

"I'm a firearms instructor, and I find that gallery loads are a great way to 'softly' introduce women, the elderly, and youth to shooting without initial bad experiences with sound and recoil."

"I specifically sell to government agencies, and have more work than I can keep up with."

"I serve handgun hunters worldwide."

"I started this as a niche, and now it's my main line of gunsmithing and reloading business."

"I offer intro classes aimed at converting gun banners. The first time I get a smile, I know the attitudes are changing."

"I load for everything in the areas of re-enactment, from cowboy to civil war, fast draw, Wild West shows... you name it. My "day job" is as a Hollywood armorer (don't worry, I live in NV), and my efforts have changed the minds of a large number of movie icons who were previously anti gun except in their films."

"My passion is combining horses and dogs with guns, from cowboy shooting to hunting, and my loads never spook the animals."

"I'm an outfitter, and gallery loading has made me stand head and shoulders above my competitors, thank you Smith team!"

"I've developed and patented many inventions based on the ideas in your manual over several years. I don't sell ammo, but my tools, kits and equipment have given me a very comfortable retirement."

CHAPTER 2. DANGERS

If you read the literature and surf the web, the most common gallery warning, and misconception, is that "reducing a load below minimum starting loads can cause explosive detonation." This is also called SEE, or Secondary Explosion Effect, so far only lab documented in military cannon loads!

If you put a double base powder like Red Dot into a very tightly contained, sturdy, sealed metal container, then ignite it with a fuse, you've created what's known as: a pipe bomb. Yes, smokeless powder WILL detonate (ie: deflagrate to supersonic speed, and create a detonation wave, rather than just burn rapidly), under certain conditions. Those conditions are extremely rare, and require very specific steps. After TRYING to detonate smokeless powders in small quantities for decades, we can say confidently that it is (laboratory) impossible to do so in ammo using published load data, WITH DACRON, unless you are TRYING to make a pipe bomb, rather than gallery or reduced load ammo. ***Without filler/ Dacron, the bullet itself can, on RARE occasions, and seldom in a revolver, perform a unique start-stop type of obturation that mimics a pipe bomb, due to a non-indexed flash path and flame front-- not a good thing. In your EC, you lose a test gun. Without your EC, you lose your eyesight, a limb or your life.***

Remember, smokeless powder is classified as an accelerant, not an explosive, for good reasons. On the rare occasions where (still undocumented and unproven) stories of detonation have been told, they always include a partially filled case with the powder up near the bullet, or dispersed across the shell (by tipping the gun down). This is not an issue if you use Dacron/Pufflon fill (or formula 47/Super

Grex etc. in your shotshells, being mindful of the pressure increases of using fill).

Be CAREFUL of "good old boy" web stories about Cream of Wheat/ Cereal/Flour, etc.-- they can work if you load/shoot within minutes, **but if you keep your loads on a shelf, they can clump and solidify, and create dangerous pressure/ SD spikes, depending on the humidity of your storage room and length of storage.** Black powder shooters know all about fills, wads and caps, just as shotgun loaders do, and many cowboy shooters instinctively tip their rifle up before shooting to move the powder against the primer. MUCH safer to just use fill and/or a 1/4" (thick) Dacron tuft, as long as you know the pressure characteristics of your fill and the crimp/no crimp rules!

One **VERY IMPORTANT WARNING: NEVER** use your reduced load shells/ hulls for regular loading after they've been fired. More on this in the engineering chapter, but in short, reduced loads can create unique resizing issues and affect headspace-- not an issue for additional reduced loads, but potentially dangerous with even normal starting loads.

Later, in the load and powder chapter, and Chapter 12, we'll get much more deeply into powder types. In this manual, we have created a classification system based on many variables, and decades of research. We call these classes **SRL (Slightly Reduced Loads-- 80 to 90%), MRL (Medium Reduced Loads-- to 50%), URL (Ultra Reduced Loads-- below 50%), and DRL (Dangerous Reduced Loads)**. We'll get into "ideal" reduced load powders below, such as Unique, Red Dot, SR4759, Titewad, etc., and indicate which loads had both tame pressure traces and SD's below 30, But for this chapter, we'll just point to ***a few known "culprits" that can be very bad:***

1. **Win 296**, although almost exactly the same speed as 4759, is a **very dangerous powder for reduced loads**

(below 10%). It's ok for SRL, but becomes dangerous and DRL for any others. This gives you a hint at the science: burn speed isn't everything, so DON'T base your choices on speed alone! Of course, along with H110, 296 is still considered an optimal (full charge) and proven powder for .410, even in the face of some tough new competition from Lil' Gun and Alliant 410. Remember that 410 loads are in the range of 11 to 13 grains to begin with, not a lot of volume for their objectives compared to rifle loads. Depending on whether your hulls are compression formed, experiences vary on how hard these are on hulls. Lil burns hot, 410 a little less so, and loaders typically get 5 to 10 loads out of a 410 hull, depending on burn through. Our experience is much longer life with 410 and 296 compared to Lil Gun.

Some loaders choose on an entirely different basis: ease of **metering**, but with reduced loads we **STRONGLY RECOMMEND that you hand weigh and funnel load**, NOT using powder drops and meters at all! The reason for this is that being a grain off in a 75 grain load is 1 percent, and in a 6 grain load it's 17%! Significant pressure differences and SD's can result from even tiny differences when they represent such a high percent of your actual load. Also, some of the best powders for reduced loads are NOT the best powders for metering. In fact, in the engineering chapter, we propose augmenting the old SD classification system with "percent SD" to more accurately reflect reduced load SD's as a percent of the load, not just absolute variance in FPS.

2. Alliant (Old Herc) **2400** is often used for reduced loads. The newer (slightly slower burning, very close to Lil Gun) **Alliant 410** also is being used more and more, and we are actively testing it right now for gallery loading. BUT 2400 can result in very messy variances (poor uniformity/ standard deviations/ pressure trace overlays), and we classify it as a DRL candidate except for light #2 lead with

proper filler. There are much better candidates for reduction, so, unless you want to spend hours with your chrony measuring SDs, avoid it.

3. Almost needless to say, NEVER use a very slow burning "rifle" powder like **50 BMG, R19/22**, etc. for reduced loads of any kind. FAST double base powders like **Red Dot and Unique; and the amazing 4759** (very stable SD's at very reduced lead loads), are far safer when properly filled.

4. **IMR and H 4895** are some of the most frequently seen reduced loads on the web and in some manuals. We've found, over many years, that there is a small "band" where 4895 works well, but it can be very squirrely (technical term for SD's all over the place, and wide variances in pressure curves), and thus potentially dangerous, in other applications outside its sweet spots. You can figure those out, or just use easier, more universal, more documented candidates like **Unique or 4759** instead. If you have 80 pounds of **Green Dot** laying around, we're not telling you to pack it in, we'll help make it work, but "having a powder laying around" is not always a good reason to try to use it outside of its best limits. OK, we'll stay in reality, in a tough economy, it's GOING to happen-- just make sure you use **published loads** if you try it. ***Check out the SD's and pressure trace ratings available ONLY in this manual.***

This is just an overview, we'll explore more later, and there are many other considerations. For example, if you have some **Red Dot**, but even more **IMR 700X**, you probably realize that the 700 is very clean burning, and can be a great choice for reduced loads, even though there are far more published for Red Dot.

One way to make a pipe bomb, in your gun, is a squib. You load so low that the first bullet is stuck in the front of the second, then fire the second. You've "doubled" your load inadvertently, as well as containing it like a pipe bomb.

Even if it doesn't rupture the gun, an unsuspected squib is what killed Brandon Lee while firing a supposed "blank" at him in the second "Crow" movie.

<u>IMPORTANT WARNING:</u> With partial /half jacketed bullets, there is a phenomenon known as a PARTIAL SQUIB! You fire the bullet, often with a reduced load, and it makes a **nice round hole in the target. Next round, the gun blows up!** What happened? Due to the start/stop of some reduced loads, the bullet hit the lands at a slight angle, and then the powder's combustion wave "indexes" the lead-- sending the lead core to the target, but leaving the jacket, or part of the jacket, jammed in the barrel! ***<u>MORAL: a hole in a target is NOT proof that you don't have a squib!</u> When in doubt, or upon hearing or seeing ANYTHING unusual, it takes only SECONDS to check that your barrel is clear of obstructions.*** CRIMPING properly (at the junction of the half jacket and lead nose) can help avoid this problem, both by allowing uniform pressure to build and keeping the core attached to the jacket. Squibs have been especially problematic in our research in .38 and .357 subsonic rounds (which of course many are) loaded to 600 FPS or lower (eg. 3.5 grains of Unique behind a 162 grain .38 spl.).

Another heads up is military surplus cases: because they are often made more thickly, pressure increases can happen due to lower volume than intended in case capacity. Some also are not appropriate for reloading, and others are brass plated steel (use that magnet to separate!).

Also remember, it takes 30,000 to 60,000 PSI to rupture many modern average rifle chambers, and less than that to rupture some cartridge shells (BUT far less to rupture some shotguns-- see .410 example below). Many cartridges are specially formulated for their caliber, and differential pressure can affect the cartridge even more than the chamber. Over HALF of factory new chambers are not uniform as well, affecting run out, wall thickness and alloy concentrations. A lot of what has been called "detonation," is really high speed escaping gas from a ruptured cartridge, which CAN destroy the action of your gun and send parts flying toward you or bystanders. Technicians might argue that escaping gas IS the definition of an explosion, but **experts in bombs and demolition science** would counter that we'd have to refine that with much higher speeds and a detonation "wave" to classify that (relatively) slow, usually subsonic gas escape as an explosion, let alone SEE.

We'll get more into the science of why this is true in the engineering and science chapters, and what to look for, what to avoid, and what to be sure of-- but something is far more important. That is the fact that **gallery loads DO explode guns**, and the 99.999% reason is double or greater charges, or charges that look "reduced" but are, in reality, too hot! When you are loading a case far below its compressed maximum capacity, it is FAR EASIER to double or triple charge a case! This brings us to our NUMBER ONE rule in gallery loading: ALWAYS LOAD REDUCED LOADS ONE AT A TIME, TO COMPLETION. ALSO GET IN THE HABIT of **tipping your case toward the ground** before each powder dump/ add/ funnel/ scoop!

In other words, DO NOT use your progressive/ turret press, and DO NOT prime a batch, fill a batch, seat a batch... do **each one** to perfection, by itself, on its own, in turn! It is impossible to double charge a load if you seat the instant you're done charging, without putting it down and doing other things, or being distracted.

Yes, you should use, 1/4" Dacron batting to take up volume, and signal fill completion at the same time, but still, do it one at a time! If you hate changing dies, get two or three hand presses, of which Lee is around $19, and use them as "progressive" single loads, still seating after each charge, one at a time. This is the only way to be sure you aren't double charging when you've got a shell too dark and large to see the powder. Corollary: **Loading Blocks are dangerous** for reduced loads, and should only be used for COMPLETED cartridges (bullets seated).

For the same load objective, isn't it better to use a higher volume powder like Blue Dot over Red Dot to avoid detonation? We'll explore the physics and chemistry of this a little bit later in the advanced sections, but most of the good (practical) advice about using higher volume powders whenever possible is for another reason than rare detonation: avoiding double charges! If you "see" the powder, the logic goes, you're less likely to put more in. True, if you're stupid or drunk, otherwise... come on! Under those conditions, you shouldn't be loading at all, whether reduced or max.

While we're on the topic of Darwin Awards-- some beginning reloaders have accidently zeroed their digital scales without taking the powder pan into account! The older balance beam scales wouldn't zero without the pan. Some digitals have integral "pans" as part of the scale, which can't be removed. Since these light little pans can still weigh 40 to 150 grains, it's unlikely that you'd miss this mistake, but if you think you're dumping 80 grains as a beginner and you're dumping 40-- you get our drift-- unintentional "reduced loads!" **Even if your powder is on a thin layer of paper when you weigh it, take the paper into account!** This is even **more critical with reduced loads**, where a couple grains can make a BIG difference in both safety and performance.

One of the "risk" measurements in any field is variance- - the standard deviations (SDs) or differences in various trials surrounding the expected values or averages of, in this case, velocity. In loading, these are seen as the resulting distribution of accuracies, pressures and velocities over a number of trial shots, and how they fit the "normal" distribution (bell curve). If you use a very low load of a very slow burning powder, you will often see terrible variance, and greatly increased risk. Of detonation? Yes, but more realistically-- of high pressures and wild inaccuracy, or of squibs.

Even if a slower burning powder formula is technically "safe," if it's inaccurate and highly variable (as seen via the SD's in your chrony with your ballistic software), it's not worth working with. Hence, **ALMOST ALL reduced load formulas use relatively fast burning powders, but the reverse isn't true-- all fast burning powders are NOT safe or uniform for reduced loads!** Think of distributions like a curve or wave dipping up and down around a central line. A "tight" curve will be predictable, safe, uniform and reliable. Wild, non-uniform swings can both dip into key holing (wild inaccuracy with end over end spin), and climb into dangerous pressure extremes.

Factory ammo SD's hover around 35 FPS across all brands and all calibers. At one standard deviation, that means you'll, on average (over say 100, not 10) shots, be plus or minus 35 FPS muzzle velocity about 2/3 of the time. We have a lot more about this in the engineering chapter, but as a rule of thumb, if your SD's are doing better in large shot tests (at least 40 shots) than factory (meaning well under 35 FPS SD-- such as 25 FPS or less), AND your bench groups are close to a couple inches at 100 yards (or a couple inches at 25 yards for a gallery pistol ball, even if consistently low-- it's consistency that matters)-- then you're working with a reliable, uniform burning load, holding other internal ballistic factors (such as rifling, forcing cone, case expansion, fit to lands, primer, wads, fill, etc.) constant. Remember, even wildcatters won't take their new loads out of the EC (Explosion Chamber) onto the range until a second lab, usually a manufacturer, has validated safety. The exception of course is published loads from manufacturers.

But, some argue (wrongly), if it is the nitroglycerine component in a fast burning double base powder that creates the detonation wave in a pipe bomb, why wouldn't this be true of accidental detonation/ SEE, in reduced loads? First, the pressure profile harmonics and gas expansion are entirely different in a closed container, and second, initiation is with a physical spark/ detonator in pipe bombs, and with a primer in shells. Contrary to some popular beliefs, primers are NOT a "spark" producing initiator. They produce a chemical heat reaction, which is still initiating subsonic deflagration (burning), NOT supersonic detonation.

This also debunks the "I use magnum primers because the spark is more likely to reach the powder if the gun is pointed down." True, it is sometimes necessary to tip a gun up with a small load of powder if we are foolish enough to risk loading without Dacron, but the magnum primer sends hot flaming particles-- heated, initiated FUEL with oxiders, sensitizers and binders, toward the powder, NOT a spark. If you're rightfully concerned about position, PLEASE use Pufflon or Dacron just like in black powder loads, but follow the manufacturer's PRESSURE warnings (yes, Pufflon, and any other fill, ADDS pressure)! Magnum primers DO work better with some reduced loads, but both magnum primers and "fill" also increase pressure-- follow the load guide on primers. A magnum primer in a partially filled case, without fill, and the powder purposely tipped away from the primer, is a detonation waiting to eventually happen, even if only once in every 100,000 trials.

Q. How come quick draw shooters don't blow themselves up all the time-- they shoot from their gun pointing down (in the holster), with an extremely reduced load and a very hot (209) primer, all the time?

A. This is a trick question-- those loads are primer only-- no powder to detonate or burn! DON'T try adding a "few grains" in this application, even with Dacron! (Note: The shooters who use black powder in these loads to pop balloons are using the BP as a projectile, NOT a propellant)!

Another caution: dangerous pressures (a far greater danger than SEE or detonation) can result if you use a crimped, **jacketed bullet with a load tested and designed for cast bullets.** The very dangerous yet popular "web" formula of a certain number of Red Dot grains for ALL reduced rifle loads is **beyond idiotic** (due to pressure, not SEE). You MUST tailor your load to the bullet type, caliber, fill, crimp and weight! In fact, crimping HELPS pressure (and avoids squibs) in some formulas, yet creates dangerous pressures in others.

If you can't stick to this with factory jacketed bullets, stick to casting instead for your gallery loading, including very effective lead balls! If you don't want to cast-- buy buck shot by the pound (**0=.320; 00=.330; 000=.360;0000=.380**). BPI's non-toxic 00 calipers exactly at .330 at the little "band" on the ball. .410 cast ball loads also come in .375-- caliper to be sure! See Chapter 13 for some pictures and loads. Nice alternative for your H and H as well! As mentioned previously, removing a jacketed squib from a barrel can be a nightmare. If you want to play with jacketed (such as a .22 varmint hollow point custom filled with grease, sintered powder, poison, etc. for expansion or quick mortality), do so AFTER you have your reduced loads uniform for that gun with lead first!

As far as the **genuinely STUPID "one size fits all" Red Dot formula** that's been on the web for years-- only ONE RULE applies: **JUST SAY NO!** SAY NO to ANY load that isn't from a PUBLISHED testing lab, or bullet/ powder seller-- ie: manufacturer!

A Nightmare Example of Stupidity

To put it simply by way of a real life example we've tested many times, **take your little single shot 45 LC/ .410**

rifle/shotgun as an example. 1/2 ounce is about 219 grains. So, you look in the book and find a formula for the LC of **7+ grains of Red Dot somewhat close to the 220 grain bullet weight of the much smaller "length" (capacity?) 45LC (handgun) shell (with a lead cast #2 bullet of course), that is subsonic.** Next, you figure that in SHOTGUN loads, you've seen **(STUPID/DANGEROUS)** "universal" loads of 13 or 14 grains of Red Dot floating around the web. Figuring that your little rifle/shotgun will easily handle that, **since you're pushing about the same amount of "weight" with 1/2 ounce LEAD bullet in a much shorter 45LC shell already, with the same gun chambered for both 45 and .410 pressures**, you reduce your load to 12 grains of Red Dot behind 1/2 ounce of LEAD shot in your .410 hull.

What happens? **YOUR GUN BLOWS UP AND YOU DIE.** We can't put it much more simply than that. It blows up for a couple reasons: 1. Shotgun loads are categorized not just by weight of shot, but also type of wad and hull (a Remington RXP 12 wad weighs 38 grains by itself!-- ALSO NOTICE that by calling it a 1-1/8 oz wad, Rem is talking about shot capacity, NOT weight of the wad itself!). 2. Most importantly, the Red Dot formula you came up with on your own (or on the web) **delivers 36,680 PSI** with wads and the buckshot and fill you loaded-- **regardless of the "weight equivalence"** of 218 grain 45 LC bullet vs. 1/2 ounce of lead shot.

Your little 45LC/.410 **shotgun/rifle is rated for only 16,000 PSI at the breech/barrel juncture**. **BOOM**. A "reduced load" is not the cause of this terrible accident, it's actually an **OVER MAX, FAR OVER PRESSURE LOAD that ONLY LOOKED REDUCED**. What seems like a "big" load of 70 plus grains in a 375 H and H DOES NOT compare to 6 vs. 12 grains of Red Dot in pistols and shotguns-- they are as different as squirrels and water buffalo in PSI tolerance and grains at max!

(Note: we shouldn't say this, but most newer 45/410s ARE designed to the SAAMI specs of the higher pressure caliber with a bit of margin, which is a bit higher (reflecting the 14,000 of the Colt instead of the 11,000 of the .410), but there ARE many older guns, especially early shotguns, that are lower, and with such a relatively small volume as the 45/.410, 14 vs. 7 is a 100% **(doubled) increase** in load, a percent increase which WILL destroy most guns). The problem is: **there is no way of knowing for sure**, because manufacturers WON'T RELEASE real pressure numbers, they will only say "follows SAAMI specs" for the sake of their own liability.

For example, the average 44 magnum is rated at about 36,000 PSI, so the "Red Dot mistake" above would potentially even destroy a hefty modern 44 mag! Since 45 Long Colt has a SAAMI spec of only 14,000 PSI-- well, you see the problem! Even the venerable 45-70 RIFLE (and most rifles are above 40,000) is only 28,000, and would have been destroyed with that load.

A great general safety rule is: <u>if you're using a shotgun powder like Red/Green/Blue dot, or a fast pistol powder, use it with your cast balls and bullets in your handguns and rifles, and use it with only published loads for those calibers and guns.</u> If you're trying to gallery/ reduce load a factory jacketed bullet, use the published reduced load formulas of other slightly slower (but still fast) rifle powders, NOT Red Dot, unless you have a specifically published, SAAMI standard tested load.

In a high bullet grain (350 to 400) 45-70, for example, this can really make a big difference in powder grains, as in **13 grains of Unique vs. 28 grains of SR 4759** (both using Dacron fill, both published loads). Since 45-70 is a favorite for reduced loads, we include these warnings since it's pressure rated below a .357 pistol, even though it's technically a rifle caliber. 4198 is another good option for reduced loads. (4895 is highly promoted for reduced loads, but far less effective and much more dangerous/ non-uniform in the sense of variance, than others we tested, even in published loads).

Of course, black powder loaders know that to push the same weight at just under sound barrier speed will take over 60 grains of FFg! No worry about case space with those black powder guys! Crush away, but be careful, black powder IS an explosive! (But yes, also crimp, so your bullet doesn't "grow!"). Speaking of growing, some loads exceed SAAMI maximum cartridge OAL by design, and are therefore intended ONLY for single shots, not trap door or lever action rifles. (We mention this because we know you lever guys love 45-70).

Speaking of black powder guys, fill cards, and Dacron-- you can buy expensive and messy Pufflon for fill, or use a trick from the old gallery guys for your Dacron-- get "low loft" (1/4" THICK) Dacron (or "polyester") batting at Hobby Lobby or Joanne's Fabrics in the fabric section of the store (see pictures). This is intended for filling blankets, toys, comforters, etc. Don't get coated/ fire retardant, treated fill (for obvious reasons-- we WANT it to burn), just get the cheap stuff. **A few bucks worth will last you a lifetime.** The general rule is about a 5/8" square weighing about half a grain, 1/4" thick, although you can vary this slightly with your case space. Realize that **pressures do increase, sometimes dangerously, with <u>anything</u> you add to the shell!** In defense of the Pufflon guys, the stuff does lubricate your barrel and action, and is really clean, with a 100% burn, unlike the folks who use TP and get a shower of confetti with every shot.

Don't forget, there is a difference between "fill" that is Dacron holding a reduced powder load in place, and "fill" that mixes with shot in a shotgun load! Shotgun fill has to be measured as carefully as powder. Also, when you hear about "wads" -- in shotgun speak, we're usually talking about the plastic "wad" which covers/gas checks the powder on one end, shock absorbs in the middle, and holds the shot (shot cup) at the front end. To confuse you more, don't mistake that kind of "wad" for the "base wad", which is part of the manufacture of the hull, on the metal end near the primer/hull base.

Another safety issue on burn speed. NEVER try to substitute either "close sounding" powder numbers, or "close burn speed" powders for each other! The COMBINATION of caliber, bullet weight and type, and case configuration determines pressure, NOT the powder alone. Very similar burn speed, or name sounding powders, can have HUGE differences in actual pressure. For example, H-4895 and IMR 4895 are quite similar, whereas H-4831 and IMR 4831 are very different (H is a bit slower, for one, although **BOTH 4831's are way too slow** for effective and safe reduced loads). Offhand substitution can destroy your gun, or your hand, vision or life, way faster than some unlikely SEE event.

If you want an example, just check your charts. In one bullet, Unique will be closest to Red Dot. In others, it will be closer to Green or Blue in charge grains. **It's NOT the powder, it's the specific combination that matters**, which is why we **HAVE to use the published bullet and powder manufacturer's charts** and NEVER substitute powders as if they were "similar." At Smith Tactical, we have many proprietary gallery loads that are only published under specific LE or Military contracts. Although the SD's printed here are from thousands of our own tests, the loads are strictly from published data. If an SD didn't pass muster (at least in the mid 30's, per factory loads), we simply DID NOT include it, even if it WAS published elsewhere. **Only 3 Grains of Red Dot? See Chapter 6**!

You might have heard of some shooters getting "consistent" SD's in the single digits. Our SD's are based on HUNDREDS of shots, not a dozen! We've found that consistency means evening out, and if someone has a load that is a supposed 9 SD, it's often not repeatable (meaning it isn't a statistically valid SD due to too small a sample size). **Minimum ANOVA fit standards tend upward of 80 to 100 shots PER LOAD. This is the reason you might not find one of your "pet reduced loads" here-- to save you time and money experimenting, we've left out the squirrely high SD loads, even if they are published loads!**

The testing of those custom, commissioned loads at Smith is far more extensive than any bullet or powder manufacturer. Yet, when we want to make one available to the reloading public, we always submit it through a powder or bullet manufacturer, or go back to the published charts! This gives double or triple assurance that the load is safe. **You WILL find formulas in this manual that you won't find anywhere else**, but these are not our proprietary loads, they are from our extensive database of proven loads, many out of print, and no longer available except to manufacturers, but all from published and tested bullet and powder makers, equipment and accessory suppliers, or testing labs.

If you're determined to experiment in load development, invest in an explosion proof chamber and remote trigger, and test fire your loads remotely! But don't forget variance-- if you're not seeing sub-25 FPS consistency on your chrony and in your groups, that "next" load might be the one that demolishes a gun and an arm, life or eyes. We don't take a load out of the lab until the Standard Deviations are as tame as factory loads for 100 rounds-- at least 35 FPS, often MUCH lower. **The thing is-- there are enough published loads to experiment endlessly! Sticking to them will provide a lifetime of fun and safe new business, not to mention insurance rates that are**

reasonable if you're loading for LE or the Military. Even in an explosion chamber, if you're not a gun manufacturer, blowing up a prized gun is no fun.

Case Inspection

One thing we do religiously with reduced loads is inspect the spent cases. This of course assumes one knows what to look for!

Remember your basic internal ballistics: pressure causes cases to expand and seal the chamber, which keeps gas from causing mischief, damage, and fouling. If the pressure is reduced too low for that case/chamber combination, there will be a poor seal, and a resultant dirty case, chamber and possibly "ringed" action.

You can also caliper the case to note expansion vs. higher pressure loads.

One solution is to use a **custom fire formed case for your reduced loads**. Where do you get such an animal? Easy: from your normal loads! A hand loaded or factory round, once fired, is automatically "fire formed" to that action/ chamber. If you neck size instead of full length size, you've got an automatic "pre formed" sealing case. Combined with **"test seating" bullet length to your grooves** by closing the action on a lightly seated bullet to determine "custom" size for that particular chamber, these can give extreme accuracy with optimal headspace and seal, even in reduced loads-- far above generalized "one size fits all" factory loads. To crimp or not to crimp is another whole topic we'll cover below!

In addition to improper expansion, of course we'll check for cracks, bulges, splits and other signs of excessive pressure, including unusual signs such as out of round spots that signal a charge had a "shaped" character. This is not normal with proper fill, but is with very slow burning rifle powders in very large cases. The engineering chapter gets into much more technical detail on these issues, but this is enough for a good start, for now.

The corollary to case inspection is **CHAMBER INSPECTION**. Look for signs of partially burnt powder, fouling, ringing, scarring, etc., which are all signs of increased pressure and/or inconsistent load burns.

CHAPER 3. GETTING STARTED

What will it take to become a successful gallery loader, research or ballistics lab? That depends on your objectives: do you want to load only occasionally for yourself and friends, or do you want to make gallery loading a new business or career? What about liability if you go outside your own use?

Check out the pictures in Chapter 13, as they give a range of the various levels of tools and equipment in which you can invest, from **almost no expense, to a complete workshop and explosives/ ballistics/ supplies lab. In these days of ammo bans and restrictions, and skyrocketing ammo prices (due often to a demolished dollar value as well as bans), even the venerable little .410 can be reloaded in a way that pays for all equipment with one weekend run!**

For example, the simplest gallery load offers something no other load can claim: **a completely tool free load**. You pick up a pre-drilled 209 shell, snap a primer in with your fingers, pick up a wax bullet, and seat it in the shell with your fingers. Bingo-- hand loaded shell, literally!

See the pictures for the next simplest method as well-- the basic tools used for decades before dies and presses. These include dowel rods, rubber mallets and nails. (For decapping, not seating a primer. Do we need to tell you not to hit an unfired primer with a nail? No Darwin awards here, please). Sure, decappers and hand priming tools are nice, but don't forget that loaders used much simpler methods for over a century.

At the other extreme, one can completely outfit a combination gunsmith/ reloading shop specifically designed for gallery loads. This can even include custom guns made to shoot certain loads accurately. At low pressures, resizing is often unnecessary (eg: primer only loads). Yet, gallery loading can also involve equipment not normally found in the average hand loader's shop, such as casting dies, neck sizing dies for fire formed cases, lead ovens and pots, decibel meters, plastic and wax making molds and equipment, paint and dye marking technologies, Dacron, Cerrosafe and other bore calibrating compounds, and additional items more often found in black powder or casting shops. Neck sizing isn't just for accuracy-- in reduced loads you can have headspace problems due to incomplete shell expansion. For this reason, **NEVER use a reduced load shell for normal loading after it's been fired-- mark it ONLY for reduced loads for the rest of its life!**

It's also almost impossible to find a swage press today, for example, as many manufacturers have gone out of business due to lack of demand. You'll have to work with Lee or Lyman to get a swage die that will work in your normal hand, cast or progressive press, or buy one of the Lyman casting kits, some of which still come with swage presses. (Swage presses and dies help form lead bullets to the proper size after casting. Many loads work fine right after casting without the need for swaging-- depends on your casting die, application and barrel grooves).

When you get to chapter 5, on choosing a niche, your strategy will also drive the tactics of which tools and equipment you will need. For example, if you decide to specialize in 50BMG reduced loads, a very important, rare and growing market today, you'll need a specialized 50 press (both expensive machined presses, and inexpensive cast presses are available), because 50 dies won't fit normal presses. Rule of thumb: if you're loading for reduced/ gallery, the cast press is fine; if it's for extreme sniper accuracy, you'll want an expensive machined press, and possibly specialty components and tools.

We're not going to hype any given manufacturer, but you can't go wrong with RCBS, Lee, Lyman, Dylan and Hornady, and MEC is the "king" of shotgun loading for price/ value. Of course the fragile little Lee Load All is ok for occasional shotgun loads too, especially the kind you'll be doing reduced, rather than the higher volume, strongly built quantity work on a MEC-- but even the single stage MEC is not that expensive compared to its brothers and sisters in the cartridge rather than hull world. "Single stage" in the shotgun world means pulling one lever and moving the shell by hand from place to place under the descending presses, rather than changing dies for each step. Sinclair, Forster, Lynwood Harrell, PPC, Stoney Point, Hart and Wilson are less known classic precision tool, die and gauge makers.

If you review the types of gallery loading, and the classes, you'll easily see there are endless possibilities for specialization. You can specialize by customer, by caliber, by bullet type, by gun type... and on and on. The Niche chapter below will warn you about trying to be all things to all people, but at a minimum, even if you specialize in, say 30-30, you will need a variety of powders, shells, primers and projectiles to be the "best of the best" in that caliber. Volume and quality will both dictate your press choice(s), but hand presses are great for getting into it, with fairly good volume, and unmatched accuracy, "feel," and safety.

You'll also have to familiarize yourself with the potential firearms that you'll be loading for, from bolt and single shot to lever and pistols. The T-C 30-30 can be one of the most successful gallery platforms, as can the 45-70 BFR.

Customizing a projectile to a gallery platform gun requires slugging the barrel, and being very aware of chamber dimensions. Reduced loads can mean reduced accuracy--UNLESS you compensate with more care in your designs. Factory loads, for example, are made for the "masses" -- the minimum average dimension that will fit the tightest chamber out there. Because of this, they are like the three bears -- too big, too small, and seldom just right.

If they make up for this slop with hot loading-- you can reverse engineer the process. Simply slug the barrel (stick a cotton plug in, melt some Cerrosafe, remove the plug, measure the lands/grooves), and gauge the chamber. You can gauge the chamber without investing in expensive go/ no go headspace gauges by simply inserting a partially seated uncharged round and closing the breech to see where the bullet and cartridge end up size wise.

Some loaders say this doesn't work, but it has proved to be incredibly accurate in our experience, even compared to completely slugging the chamber. What you've now done is beat the factory guys at their game-- CUSTOMIZING your rounds for that gun, just as all match and competition shooters, and many sniper and dangerous game PH's as well. Neck sizing is a must, letting each gun "fire form" the case to its particular unique dimensions.

Was that 7.62 x 39 European or Russian standard? If you're working on a .308 or .310 assumption and the barrel is .311 (as is the CZ), you can swage your projectile to be both a perfect fit and have perfect lands contact when chambered. An alternative is to make or buy a bullet with a hollowed base-- it obturates to make up for the extra space. This works well in the poorly designed and rifled Mini 30, as well as the tack driving CZ 7.62 bolt action (which actually is .311). See the science chapter for more on another important issue with 7.62/ .308-- rifling/ twist.

If you're specializing in cowboy shooting and/ or revolvers, you'll also have cylinders and forcing cones to deal with in engineering your loads. The moral: you can make up for diminished accuracy at reduced loads by customizing your loads more closely to individual firearms. If you buy or make cast bullets, they can even be far more accurate than fully charged factory loads at the same distances!

Of course, if your objective and niche is self defense and simunition training, accuracy is no big deal, because most encounters will happen in shoot houses and at confrontation distances of 20 feet or less. Some very slow gallery loads (below 800FPS) are very accurate in terms of grouping, but groups can be consistently low, which means a simple sight adjustment. With that objective, you'll be investing more in projectiles, than in customizing rounds to guns.

You also will need the guns in which you intend to specialize too! Handgun hunters using rifle loads will have to become familiar with expensive guns like the T-C (if Smith keeps them in business, and survives themselves), the BFR, the 500 SW magnum, the 460 SW magnum, and a variety of somewhat less expensive Taurus and Ruger options. All the loading equipment you'll need for a lifetime won't cost as much as one SW 460 revolver in many cases! If you don't have a fortune to invest, make a deal with a buddy or local range to borrow their guns in trade for business or loads.

With DC and many States continually (on and off) going after ammo, supply and demand varies. It's not a bad idea to stock up on hard to find items like primers and shells, unless you're reloading from a plentiful supply of factory brass or ammo. A good supply of #2 lead and the relevant casting dies will assure you've always got some kind of projectile available. Powder also comes and goes in supply, but when stored carefully-- cool and dry-- can last many years. If it smells like acidic sulfur (bad), it's bad, if it smells like alcohol, acetone, or gasoline, it's generally fine. Check it in a non critical environment before confronting that lion, terrorist, mugger or bear, if it's over a few years old. Modern powders are made to last a lifetime IF stored in cool/dry conditions.

To be honest, **a lot of folks pick their niche based on the guns they have and like!** They start as a hobby for inexpensive practice or less noise and recoil, and sometimes move on to making cash on the side after friends see how much fun gallery shooting can be. By all means, if you eat and sleep 45-70 or .308 or 30-06 (my personal passion is 375 H&H), then do get started in those areas. Your initial fun and success will suggest additional expansion. One of our most successful customers got a 30-30 T-C, and didn't like any factory ammo-- gallery loading ended up being the unplanned result of a lot of trial and error.

If Armageddon happens while we're making other plans, our hobby can also suddenly become a survival tool, both for self defense and hunting. The ability to recycle our loads can be a life saver, regardless of formula. And investing in tangible items-- presses, tools, supplies and ammo-- will beat inflation just like gold, silver and copper, and with DC printing money like there is no tomorrow, what better place is there to invest now?

Here is list of basic, intermediate, and advanced tools and equipment for your ballistics lab/ reduced load development lab, just as a starting point:

Basic

Powder, primers, cases or hulls, bullets or shot

Hand primer

Hand press

Dies for your intended calibers

Shell holders for your intended calibers

Work bench

Scale

Funnel

Crescent wrench

Dacron fill

Cleaning supplies

Bullet puller

Loading blocks

Powder measures

Lubricants

Intermediate

Above, plus:

Neck sizer dies

Factory crimp dies

Chronograph

Use of ballistic website

Primer pocket cleaners

Case trimmers and gauges

Caliper

MEC press for shotgun handloading

Cast or forged/machined bench mounted press

Powder trickler

Powder columns

Shot columns

Tumbler and media

Advanced

Above, plus:

Casting molds

Casting oven/ pot

Lead

Lead dipper/ ladle

Specialty projectiles

Wax making equipment

Swage press or dies

Ballistic Software

High Accuracy specialty dies

Pressure Tracing hardware and software

Specialty presses

Case milling machines

Drill press

Explosives lab components and EC

Ballistic Microscope

Decibel Meter

Supplies/ accessories manufacturing equipment and tools

MUCH MORE! Advanced Gallery Loaders have a whole slew of custom fillers, tools, supplies and accessories we'll cover when we get to loading.

<u>One Key Tool: YOUR PRESSURE SIGN</u>

Put up a sign near your EC, in your ballistics lab, that gives a few telltale signs of overpressure, and look at it frequently:

1. Blown out primers that fall out after firing

2. Shorter case lives

3. Extraction difficulty with cases after firing

4. Enlarged primer pockets ala #1 above

5. Marks on the case head from ejection/extraction

6. Primer pitting and cratering-- pressure flattens/smoothes primers

7. Gas leakage near primer

8. Bolt face pitting

9. Expansion of case heads

There are many more, but these will get you in the good habit of checking!

CHAPTER 4. A LITTLE SCIENCE

This is your "accuracy" and sniper chapter (more on this in the Engineering Chapter too), plus some high accuracy candy about **optics, rifling and sighting**. There is a lot of false info, especially on the web, about accuracy, sniper loads, and ballistics, **especially when it comes to reduced loads**. To understand both reduced load performance and accuracy, we also ***have to understand a little more about ballistic curves and scope sighting,*** from the sniper or hunter's perspective.

The "best" accuracy is the **accuracy that gets the job done.** If your simunition™ loads can't hit a metal pie pan at 100 yards, but can accurately simulate combat in a shoot house, you've done your job, and not over-engineered your loads! Additional muzzle velocity or aerodynamic spitzer shapes (high BC's) don't add a thing. Most primer only wax bullet loads always shoot low, but still can have great groups. Just aim high, or adjust your sights!

The previous chapters touched a little on internal ballistics (powder, primer, action, bullet, case, etc.), and this one will add a bit about external ballistics (flight path of the bullet from muzzle to target) and terminal (impact characteristics) ballistics. We'll also cover a little about rifling, which is actually internal ballistics, as well.

As a general rule, the higher your muzzle velocity, the flatter your flight path and greater your recoil, but this also is affected by other considerations, such as bullet shape and weight, stock design, and muzzle accessories. **It's sad that ballistic software has caused most of the manual producers to stop publishing ballistic tables and charts.** Sad, because these are great visual aids to understanding the real essence of accuracy.

To begin with, there are two basic variables that used to be the "entry point" for your chart predictions and calculations-- the BC or Ballistic Coefficient of the bullet, and MV or muzzle velocity. BC is, in non math terms, the ratio between the **bullet's sectional density (mass/caliber) and its shape ("pointiness" or sleekness, or aerodynamic shape)**, or the "form factor" (called "i"). The higher the BC, the better able the projectile is to overcome aerodynamic drag-- or air friction in plain language.

A BC of .1 compared to a BC of .5 ISN'T necessarily BETTER-- it just means that the first probably has a rounded nose or draggy caliber and the latter is a spitzer, or pointy projectile better able to overcome air resistance. As we mentioned, a pointy projectile also is less able to overcome brush deflection, and often less able to retain weight-- which is why a relatively "lower" BC in a round nosed 416 Rigby is preferred for **dangerous game at shorter distances** than a 300 win mag spitzer. The "BC" advantage of a long flight path is not an advantage there, especially if the jacket separates and the bullet **loses most of its weight when it hits thick skin** (bonded or not).

If you enter info into today's software, you're often starting with chrony (velocity) info. In the older tables, you simply cross referenced the **muzzle velocity with the BC** of your projectile, and got a path chart between 0 and 1,000 yards showing rise, zero, and drop. Remember, these measurements are based on line of sight, usually through a scope, and if your scope is mounted higher than the usual 1.5 inches for hunting/ sniping (low mounts), the numbers WILL change. A bullet doesn't really ever "rise" relative to the barrel-- the scope is always pointing somewhat down, so you're always subtly pointing the muzzle/barrel up a little.

When the path "rises," it is always calculated relative to your line of sight, and that of your scope or sights. Some advanced scopes "run out of clicks" for far distance weapons (eg: 50 BMGs at 1500 yards), and you can actually buy "20 degree down" mounting rails for your scope, which are milled to point the scope a little more downward from the get go, so you don't run out of clicks.

The "zero" of a path is simply the place where the bullet path and line of sight intersect, without the bullet having a rise or fall at that point. Most rifles are zeroed at 100, 200 or 300 yards, with 100 being the most frequent for hunting, and 200 for long distance sniping. **A very practical technique used by hunters and snipers is to calculate what's called the MPBR**, or Maximum Point Blank Range. This is actually just the **OPPOSITE of what it sounds like**-- it's not "point blank" or right at the muzzle, but is actually the FARTHEST YARDAGE you can shoot without exceeding a certain ACCURACY BRACKET. For varmints, that might be 1.5 inches. For humans and big game, it might be 3 inches.

If your Ballistic Coefficient (.483) came from a .308 180 grain spire point, and your MV was 2,600 FPS, the chart will show that to stay **bracketed** at 3 inches **(no more than 3" high at any point in the trajectory, and no more than 3" low at MPBR)**, you'll be about 2.89" high at 100 yards, and your maximum range to go no more than 3" low (MPBR) will be 259 yards. Your "mid range trajectory" (the point where you hit 3" high) will be 135 yards, and your actual zero will be at about 229 yards. This technique is far better than "zeroing" at a given yardage, then continually adjusting the scope for distance, as it **guarantees a kill zone on a wide range of average shots and multiple distances, without a lot of adjustments.**

Bore and live sighting

Below are several articles we previously published on the gallery site and e-versions of this manual, which give a flavor for how sighting relates to accuracy. These are VERY IMPORTANT to understand if you are designing custom loads for accuracy as well as sound and recoil. ***You NEED TO KNOW how these details work if you are starting a gunsmithing or reloading business, or even a ballistics lab!***

Bore Sighting and Sighting in the Thompson Center Encore

This is a report on sighting in a new Thompson Center Encore Pistol in .308 Winchester. We loaded a number of both full and reduced, and one "CIA" assassin's load, and then tested the Encore's performance with the factory scope.

The "full" load was a Hornady 30 Cal 150 grain SST with a BC of about .371. The "reduced" load was a Nosler 150 grain spitzer. The "CIA assassin's" load was a "0" buck (.320) ball sized to .310. All three were loaded over Winchester factory .308 brass with large rifle primers. **The full load was over 43 grains of Varget. The reduced loads were over 12.5 grains of Red Dot, 25 grains of IMR 4895, 10 grains of IMR 4227, 11.5 grains of Unique, 19 grains of IMR SR 4759, and 22 grains of 4198.** The "CIA" load was over several grains of Red Dot.

The old "general rule of thumb" (**BUT-- ALWAYS only use published loads!**) to avoid dangerous pressures or rare detonation at reduced loads, is to not go below half of a fast pistol or shotgun powder's maximum (Unique or Red Dot), and not below 2/3 of a slower burning rifle powder's maximum. Of course, as we've seen above, "above maximum" is also just as easy to reach, or easier, if we don't consult our load books.
Reduced loads are popular in Thompson Center pistols to tame the often hefty recoil in cartridges like the .308, with barrels as short as the present 15". Some gunsmiths and reloaders consider using red dot or unique an unproven "wildcat" technique (adapting essentially shotgun powder to rifle loads), but this is not true: the cast bullet and

reloading catalogs, as well as the loads listed below, have had reduced load formulas for this combination for years! The web makes this looks innovative and mysterious; cast bullet loaders have known about it for years.

We first used the Leopold "Zero Point" magnetic bore sighter. This is technically known as a collimator. It is different, and far better, than the eariler Leopold or Bushnell magnetic bore sighters, because you can see through it to the target. This allows ONE SHOT adjustment. All these collimators are set for 25 yards "on the paper," which is the first ballistic crossover point for a 100 yard zero, but "on the paper" at 25 is NOT on the paper at 100!

For this reason, the very cool Zero Point's see through feature comes in handy. You sight the scope at collimator zero overlay, shoot one round at 25 yards, hold the collimator indicator over the target's bullseye, then readjust the scope for the actual impact hole (moving the collimator image to the actual bullet impact point at 25 yards--- "down" in scope clicks moves the "target" (collimator image) DOWN, "right" moves the image right). This gets you on the paper at 100 yards! All this fuss is to save a lot of expensive ammo getting you to zero.

Hold overs and unders are moot with the TC scope, as it has no mil or moa or inch dots in the reticle, so you've got to go by "gopher heights" etc. if you want to reach out. For pistol hunting, the humane rule of thumb is to consider it a 100 yard or less platform anyway, especially with reduced loads, some of which are 100 FOOT loads, not 100 yards!

At maximum loads, you run the risk of both developing a serious flinch, and blowing most of your propellant's flash out the end of the flame-throwing crown anyway-- cool looking at night, but worthless otherwise. Hold unders at less than 100 yards are only a couple inches, so unless you're shooting mice, it shouldn't matter that much if you have a good 100 yard zero.

The "CIA" load is based on an assassin's gun platform using the T-C Encore Pistol. The "O" (.310) buck shot pellet is propelled only by primer and a few grains of fast burning powder, in a .308 cartridge, crimped mid diameter. This platform shoots silently and accurately at 25 yards, leaves no brass behind, and the projectile has no identifying forensic ballistic identity. See Chapter 13 for some photos, and more info on explosive tipped RL's and other exotic platforms.

Smith Tactical, and their sister company at Explosive Technologies, has engineered a number of these packages and loads for various government and military entities, including several with lethal cores of fulminated mercury, bactratoxin, and others we won't mention. See chapter 13 for some examples.

For practical plinking (sans poisons), this is a fun load that costs nothing for "bullets." A 5 pound bag of 0 (.320) or 00 (.330) or 000 (.360) or 0000 (.380) is about 20 bucks, and you can get about 18,000 rounds out of an 8 pound keg of Red Dot! Now, don't start thinking you've found a source for the tough to get .380 for your LCP!*

Also, 18,000 rounds would still have the little issue of brass and primers, even if you have the lead and powder! At this tiny load, barrel leading with the 0 buck is a non issue, even if you don't "mold your own" normally. It's still a great little squirrel gun, and a lot cheaper than $2X a box 17 HMR varmint rounds!

All the above loads brought in MOA groups (1") at 100 yards from the bench, as did the CIA at 25 yards. The T-C scope was adequate, but we'd recommend a field rest or bench rest, or monopod, as even with their generous eye relief and tough recoil profile, these scopes bobble wildly at anything over 50 feet. And that's at 2.5X, at 7X-- forget freehand shooting! Remember: hold the GRIP TIGHT with both hands in any T/C! Do NOT shoot holding the front barrel piece with one hand, or you'll have a permanent hammer mark in your forehead with the 500 SW Mag, .308, etc. **Rule one in Encore and Contender Pistols: HOLD ON TIGHT!**

**The process for successfully implementing the CIA load requires a few "casting" techniques. First, slug the chamber and barrel of your gun, such as the TC here. The best method* *is to use Cerrosafe with a little wad of cotton. .309 is a good estimate, but it's far better to actually slug, so you see clearly lands and grooves, and can then micrometer check the ball. Next, size the original "0" ball with your Lyman sizer (or a dremel, carefully, with lubricant, keeping it round!). Some folks laugh at round balls, but they are surprisingly accurate, and even in the Civil War, the "Minie" would tumble if a little out of spec, where the ball wouldn't. In any case, this is a 50 yard load, not sniping! It is quieter than an airgun (sounds like the single clap of your hands), and you can easily shoot one hole groups at 20 yards. Rats, skunks, squirrels beware, even in the city where neighbors would call the cops on any louder round! Black ops types use this load with a small custom .308 derringer that makes their mission silent, without the need for a silencer or large weapon.*

Project DT 006: Live Sight Steyr 50 BMG

Smith Tactical was presented with a factory new Steyr 50 BMG Heavy Sniper Rifle and asked by the customer to install High 50 BMG rings, a Leupold Mark 4 Sniper Scope, and then bore sight, live sight, and test the weapon system for accuracy and function, polishing the bolt and accurizing the trigger.

On the initial test, we calibrated the Mark 4 (2nd focal plane/ mil dot reticle system) at 40 and 300 yard zeroes, then final MPB zero of 400 yards (actual 500), actual MPB at 450 yards, and max range (mil dot/2 meter target) at 1050 yards, on a gun "rated" at 2000 yards. Beyond a mile (1,700 yards), the drop and accuracy of a rifle with a 33 inch barrel are too variable to consider the system *consistently* reliable, even with the Mark 4 at 25x and premium Hornady ammo, although a near 2 mile kill has been documented with this very rifle-- the exception, not the rule.

Upon bore sighting to the black, it was found that the MPB elevation was perfect with the High 50 rings/Mark 4 combination at 400 yards right out of the box, and only required a two click down/right adjustment to be dead center. This combination of rings and scope is amazing. After the initial test, the rings were re-checked for torque (initial: 65 inch pounds cross bars/15-- light firm finger/thumb-- for top screws), and did not fall within Smith Tactical tolerances (.03). These were returned, and replaced with new Badger High 50's. The new rings did not vary at all. The Mark 4 doesn't run out of clicks at this yardage, so a 20 degree down mount was not needed.

All testing was done with the Hornady 2.54" A-Max 50 BMG at 750 grains with a BC of 1.050, MV of 2,800 fps and ME of 13,000 f/p. The scope was re-set for a BC of .67 before shipment, due to the customer's ammo specs. Bullet drop at MPBR (450) with the A Max was an actual -51.69 (-11 clicks from 500 zero). Recoil was approximately 83 foot pounds. Alititude for all tests was 4470. Temperature was 70 F, BP was 29.81. Humidity was 10%, and there were no to very light cross winds.

Scope Tips

This system is now set for maximum efficiency and ease of use for a 2 yard high target at 450 yards-- the customer's requested range. From that baseline, leisure (competition) sighting should be done with a sighting/ranging scope or ranging binoculars, and elevation simply dialed into the incredible Mark 4 based on range.

Medium time frame faster sighting can be done with shooter reticle estimation of range, then dialing in correct yardage in clicks. To go from 450 yards with the customer's preferred load (just like dropping back from zero to the MPBR) will require negative clicks, approximately -15 at 300 yards, and -25 at 200 yards. Shooter reticle ranging with the (2nd plane) Mark 4 should be done at 25x. The 4 gives 1 mil for each dot in the reticle. Target range is target size in yards times 1,000, divided by mils. If a 6 foot tall target has his feet on the crosshairs and his head is on the second mil dot (at 25x), he is at 1,000 yards. With MPB ballistics set at 12 inches, +060 clicks give a consistent two foot group at 1,000 yards with the customer's ammo drop of -271.

"Hold over" ranging is not recommended with a system designed to be shot accurately at over 500 yards with a bipod or sand bags, and weighing over 30 pounds, but the Mark 4 is legendary for hold over accuracy. Since each mil as scope is currently set is about 3.5" at 100 yards, or 3.39 MOA, you can get a quick holdover estimate with this scope by multiplying the range/100 (4 for 400 yards) by the MIL INCH (NOT MOA), or 4 x 3.5 = 14 inches (for example), which is the "dot to dot" value for each mil through the scope. The "technique" is to take the bullet drop of the round you're using, and hold over the number of mils that dividing the bullet drop by the 14 inch mil value would give you. A drop of 42 inches at a bit over 400 yards with the Hornady A-Max would give you a calculated HO of 4 dots with this method, as an example. Of course this is not necessary with this system at this range, as the scope was already zeroed at 500 and set to an actual MPB with predetermined clicks (-11) at a 12 inch allowable target size. For reference when MOA adjusting, the Mark 4 has 1/4 click MOA adjustments. Practically, estimating "air" at ranges over 300 yards is iffy at best, unless you have a natural "talent". As a fighter pilot, our customer for this system could probably do it!

Scope was removed for shipping, rings left on. Eye relief was set at 5 inches for safety (maximum) due to recoil (exercise CAUTION if sighting uphill!). Leupold touts its "long" eye relief (between 3 and 5") but seasoned pistol hunters know that's about the minimum. Rings have been set so customer can re-set scope to the max forward position that is comfortable for his cheek/vision and shoulder placement (our SWAT and military customers even request mounting the rear ring on the exposed eyepiece threads occasionally, which is within spec for this scope), then add/remove the scope quickly for field sorties. The precision-machined top mount screws on the high 50 are specified at 15 in/lbs, which is practically light to firm finger/thumb tight using the Leupold wrench. Most shooters don't carry i/p gauges in the field. Recoil dictates conservatism, but the high 50's are forgiving and very precisely machined. Height was determined as the minimum necessary to clear bolt, fit the 1913 pic rail, not interfere with ejection, and yet remain as low as possible to the barrel within these constraints.

Performance

This rifle proved to be quite accurate, even at 1,000 yards. Move over Quigley. 4" and tighter groups were consistently achieved at 400 yards, and 18" groups were achieved from time to time with some tinkering at 1,500+. Most good rifle marksmen cannot hit a standing Elk consistently with an 18" target zone over 300 yards holding at shoulder (we're not talking BR shooting here!). Even with a huge 50mm scope, it is hard to quickly acquire a jeep at 1,000 yards, let alone a little piece of paper. Given that this round is intended for stopped materiel, not moving targets, the larger "engine target" size makes real mile shots possible. Don't forget that the "mile" shots we often hear about are with guns having 4 foot or longer barrels. This can make a difference of 10 inches in accuracy over a 28 inch barrel (the Steyr is 33), but you're then lugging around a 60 pound rifle. This amazing Steyr is quite *light by comparison* at 28.5 pounds. **ALSO SEE THE ENGINEERING CHAPTER for important information on break in and copper in this and other high accuracy guns.**

Penetration with armor piercing rounds is calculated at 1.5 inches of steel (with 700 military ammo), although we did not use armor piercing rounds for this test. MOA fine tuning for windage and elevation, from those seen in the targets below, were then made using traditional center-group measurement click adjustments to "move the group".

Although this is advertised as a "sub MOA" gun at 1,000 yards, we found this unrealistic if you require consistency and speed in wind and rough terrain. That would mean your groups are CONSISTENTLY tighter than the center of the 10 ring of a 1,000 yard competition target (under 10 inch groups at 1,000-- 10 ring is 20 inches). Sure, the world record is 3 or 4 inches, but with several "bench" series, the best we could do is 18 inches *with consistency (SD's below 10)*, which was measuring bullet shape and wind more than our skill! We're curious how some of these studies measure "sub MOA" in a round that pokes a half inch hole with every shot anyway! Certainly MOA measurements below 300 yards become part hype. If you're allowed to shoot 10 3-shot groups, and one is at 6 inches at 1,000 yards, and that is your definition of sub-MOA, then yes, this gun MORE than qualifies. Hand loading by case forming and fitting, and proper copper break in, will easily make this weapon sub-MOA, as we have done many times with other 50's, such as the magnificently manufactured Noreen.

SD's for both the factory ammo ranged from 23 to 33 FPS--respectable as a small percent of this hot ammo. Our hand loads came in at under 8 FPS.

The "down a jet in flight" mystique of the 50 BMG is also hype, as the A-max has a maximum altitude of about 16,500 feet straight up, with no penetrating power at that altitude. And the odds of actually hitting something moving that fast are about like trying to take out a satellite in geostationary orbit with your .22. These myths are perpetrated by gun banners, so reporters beware! CNN was actually gullible enough to fire a Barrett through a jet door at about 30 yards, and "extrapolate" from there. **Don't tell them that Weatherby makes at least 6 rifles more practical in weight, recoil and caliber than Barrett if you really want to harass a Hamas jet on takeoff.**

Little known fact: The 50 BMG is actually a scaled up 30-06. John Browning liked the ballistics and profile of the venerable 30-06 so much that he designed his "jeep killer" with the same ratios! The 50 BMG is still the oldest and most widespread military caliber in the world. Wildcatters are encouraged to continue development of the range of ballistic and design options, as this is the surest way to defeat the banners. Thanks DT, forever and always, and God bless you and yours.

A curious and interesting side note on shipping this monster: the very substantial Austrian crate is out of spec for USPS, not insurable at Fedex, and contents require "inspection" at UPS. How do you remove the steel bands, "show" the contents to UPS for insurance, then re-band with your local

UPS center staff (and a few other curious customers) standing around agape? Answer is DT's secret!

More on Zeroing Your Gun...

Zeroing relates to all three aspects of ballistics: internal, external, and terminal. Although you can zero both iron sights and scopes, this section will concentrate on scopes. When you mount a scope on a gun, imagine a line extending straight forward to infinity from the scope. Now, imagine another line extending to infinity from the bore (inside of the barrel of the gun). The scope line always "points down" below the bore line at some point, and the bullet trajectory curves between the two.

The scope line is obviously above the bore line at the beginning, even though usually tilted down, since the scope is mounted above the barrel in most cases. Ballistics show that each bullet travels in a parabolic arc of various shapes and slopes (depending on the bullet selected), "rising" above the line of sight, reaching a peak, then falling, all due to the effect of gravity on the projectile vs. inertia, drag and weight. If your "site line" on the scope is pointed down a bit, the bullet obviously crosses that line in two places-- on its way up (called the near zero or point blank range/ PBR), and on its way down (the far zero, or "true zero"). From that simple standpoint, zero means the bullet is hitting where you're aiming!

Contrary to popular belief, due to the laws of gravity, a bullet never really "rises" above the bore line; the "rise" is above the sight line of the scope. The exact definition of a zero is the point at which there is NO RISE and NO DROP relative to the two lines, and the two lines intersect-- hitting what you're looking at. If you just put a scope on without adjustment, and it's perfectly straight, and is 3" above the bore, the bullet will hit 3 inches low close up, because the bore is that much lower than the scope. In reality, these lines are never parallel, the scope is always tilted down as you turn the dial, to intersect your target with the bullet trajectory relative to the (often elevated) bore line.

Most people think of zero as the point where your bullet will actually hit the target (noun) when you aim at it, which also might be true, if you "zero" (verb) the scope to that point! The old process of zeroing was to go out to the range, look at the target through the scope (or bore!), start firing, and adjust the scope up or down, left or right, until your shots hit the bulls eye.

If your target was at 100 yards, your gun (actually your scope) was then "zeroed" for 100 yards. Since the bullet travels in an arc, remember that there was also a "near" zero, at about 25 to 50 yards for most bullet weights and ballistic coefficients (weight/shape ratio of the bullet). Since your bullet always goes up, then down, the point where you "zero" your scope will create a division between this rise and fall.

For example, if you zero a 50 BMG M2 military rifle at 1,000 yards using a 709 grain bullet (M2 FMJ-BT 2850 FPS, source: ballistic software run on Quickload), the bullet will be 92.9 inches "high" (rise) at 500 yards, and have a "drop" of 191 inches at 1300 yards.

Zeroing a single shot Noreen 50BMG at 100 yards will give "drops" for everything beyond that range-- 46.68 inches at 500 yards, and 262 inches at 1,000 for a Hornady 750 grain AMAX, for example. "Drops" in ballistic tables are given in both MOA and inches, because scopes are calibrated in both. To hit a target before or after your zero, you'll need to either adjust your scope clicks for that rise or drop, or "hold over" (aim above your target) a certain number of "scope dots" to make up for the non-zero bullet rise or drop.

Another method of zeroing is to simply remove the bolt on your rifle, look through the barrel, and adjust the sight picture of your scope to coincide with the bulls eye. Of course, your rifle will then be "zeroed" at the distance you set your target-- often 25 to 50 yards. The most modern zero techniques use bore sighters-- either lasers that protrude from the barrel, or optical collimators like the Bushnell or Leupold that magnetically attach to the crown of the barrel, creating a little "target" picture when you look through the scope.

These are guaranteed to "get you on the paper" (usually at 25-45 yards, depending on bullet, seldom at 100) without firing a shot. In the latest version of the Leupold bore site, there is a "guaranteed one shot zero" because after you are "on the paper," you can actually look through the bore sighter, adjust the scope to superimpose the previous bulls eye on where the shot actually hit, REMOVE the boresight, and shoot again.

Once your scope (gun) is zeroed at a certain yardage (typically 100 for most hunting guns, 200 to 500 for large sniper rifles like the 50 BMG, 25 for pellet guns), if you go hunting or target shooting, you will need to estimate the range to your target, then adjust your scope with windage and elevation clicks to compensate for windage and elevation differences compared to your zero. Ballistic charts tell you the rise or drop based on your bullet factors, and calculators like the milldot master can convert between distance, MOA, mils and bullet drop in seconds.

The world's best shooting competitors even use handheld computers with ballistic software to exactly compensate for everything affecting zero, from wind and humidity, to the rotation of the earth. The Cheytac has a computer built in that does the same thing in the field.

Q. Since most expensive and long range scopes are in Mil Dots, why are the adjustments in MOA, and how do they relate?

A. One strange answer is that DOD refuses to issue sniper teams range finders (even though the Bushnell 1500 is about 400 bucks!), so mil dots are still being used to estimate yardage. Since a mil target size is 36 inches high (3 feet), (as in distance between the two dots), at 1,000 yards, if a 6 foot tall man covers 2 mils in the scope, you know he's standing at 1,000 yards. Some folks ask "at what focus?" (there IS such a thing as a dumb question)-- it doesn't matter, because the reticle has the same mil readings regardless of setting. However, you should set parallax to correct yardage before zeroing, zero the first time mid-range (if 20-40X, zero at 30), and for final, at max.

MOA clicks are totally dependent on your bullet. Adjusting your scope for bullet drop from its current zero is done with MOA clicks, even with a mil dot reticle (crosshairs). If you know your target is at 700 yards (from your mil dots, or range finder), you have to "bring up" (or come up or hold over) on your target by the number of inches that the bullet falls at that distance. For convenience, let's say your bullet drops 210 inches at 700 yards. This means, from your scope zero (of 100 yards in this case), it has dropped 210/7 (700 yards, at one inch per MOA per 100, is 7), or 30 MOA. If you are going all the way from 100 to 700 on your scope, you've got to "come up" (as in, "bring the bullet path up" to your scope's line of sight/crosshairs) by 30 MOA. with 1/4 clicks, that's 120 clicks, with 1/8, it's 240.

Where do mils come in? Well, if you'd rather "hold over" your rifle with MILS instead of adjust your scope with MOA, you can remember that 1 mil = 3.438 MOA, and if your MOA come up is, say 12 MOA, the mil come up would be 12/3.438, or about 3.4 or 3.5 mils-- the number of mil dots (unless your scope has MOA dots!) you have to hold over to compensate for the drop. We chose the 12/3.4/3.4 example because it's easy to remember. Now you know both, and can handle ANY type of scope. For small targets, another "closer" 12/3.4 relationship is that at 100 yards, a 12 inch target covers about 3.4 mils of reticle in your scope. At that yardage, a 12 inch bullet drop also is about 12 MOA (actually 11.5, but close!).

If you read this stuff and ask "but doesn't that depend on the bullet?" -- you're getting to an important advanced topic-- **which factors are constant and which are variable, and which are dependent and independent variables.** This confuses a lot of shooters! The two variables that "move together" are distance and bullet drop, and they are related to the constant (math multiplication or division) relationships between MOA and mils (1/3.4 ratio).
Playing with the Mildot Master gives you a GREAT visual of how these variables interact. Once you know drop, you've already factored type of bullet into the equation, as the BC is implicit in the drop number, as well as the weight (bigger grains drop faster); weight is in the numerator of the BC

fraction (shape is in the denominator). Another name for the BC numerator is SD, or sectional density. This is "working backward" to BC-- a field technique, as opposed to checking a chart with BC to determine drop, holdover, or range.

Q. What is a "Flat Shooter?"

A. Most people talk about the 460 Smith Magnum or 375 H and H as "flat shooters." What do they mean by this? Simply: a 100 yard difference in a light .223 can mean a two foot difference in rise or drop, meaning the ballistic curve is very "curved." Flatter means less rise and fall across the whole projectile path. For example, a 375 H and H zeroed at 300 yards will only be about 5 inches high at 200 yards, a "relatively" FLAT trajectory indeed, for a caliber that is so versatile! You might argue that this is wrong, because we didn't specify bullet weight. But we tricked you! The old (1912) 375 is a mysterious caliber, that for some reason, hits near the same impact point regardless of bullet weight grains, very unique in all the world. It still ranks as "the one gun" most PH's would choose if they could have "only one gun" for all applications.

Like all calibers, it's a little the gun and a little the shooter and a LOT the load. A custom loaded .223 can take the head off a groundhog at 600 yards (we've done it). The key to cover a dime accuracy has always been custom reloading, whether gallery or full charges!

Zeroing iron sights

$D1 / R1 = D2 / R2$

For rear sight adjustments:

- **D1** is the distance between point of aim and point of impact.
- **R1** is range from front sight to target.
- **D2** is the length the rear sight must change by.
- **R2** is the distance between front and rear sights.

For front sight adjustments:

- **D1** is the distance between point of aim and point of impact.
- **R1** is range from rear sight to target.
- **D2** is the length the front sight must change by.
- **R2** is the sight radius distance between front and rear sights.

This formula calculates the *MAGNITUDE ONLY* of the sight height change; refer to the instructions above to find the correct direction for the adjustment (front or rear sight, longer or shorter). Likewise, all distances must be in the same units. That is, if a change in inches to the sight height is desired, and one is shooting on a 100 yard range, then R1 (100 yd) must be converted to inches (100 x 36 = 3600 inches) before using this distance in the equation.*An example: Consider a rifle with a distance between front and rear sights of 26.25 inches, firing on a 50 yard (1800 inches) range, with point of impact 5.3 inches too high on the target, having a front sight blade that is 0.505 inches high mounted in a dovetail. How much must the front sight blade height be changed by to fix this problem? (It will be assumed that the muzzle of the rifle intrudes into the range space for following typical gun range safety protocols, and the rear sight is hence 50 yards from the target.)*

***D2** = **R2(D1/R1)** = 26.25(5.3/ 1800) = 0.077" (magnitude of change to front sight height)* **OR M (modification)=(Miss"/Range")*Sight Radius (dist between sights)**

Since the gun is hitting too high, the front sight must be lengthened (raised) by this much (or the rear sight lowered by this much); hence, the front sight must be replaced with a blade that is 0.505" + 0.077" = 0.582" high. With this correction, the rifle will hit the desired point of impact, all other factors being equal. Or, adjust the rear downward by the same .077"

Hitting low, raise back sight or lower front

Hitting high, lower back sight or raise front

Hitting low, if you lower the front sight, when you bring the back of the gun down to meet the front with the rear, you'll be RAISING the barrel

Hitting low, if you raise the back sight, when you bring the front of the gun up to meet the rear with the front you'll also be RAISING the barrel

The general rule is the rear sight is moved in the SAME direction you wish to move the **point of impact.** Assume the point of impact was LEFT and BELOW the target. To move the point of impact to the center, move the rear sight RIGHT and UP. The front sight moves the opposite direction, so it would move LEFT and DOWN.

This might help a little with some of the Thompson Center reduced loads below, which often use iron sights!

"Hunter's Zero"

If you're bored or confused by all the mils and moas and fine adjustments, don't despair! There are a lot of rules of thumb, like the MPBR, that will serve you very well if you load accurate rounds and do just basic sighting, making every shot "close enough for dinner".

One of the easiest formulas is what has been called the "hunter's zero" -- or a few rules of thumb that can provide **VERY accurate results with minimal math.**

The hunter's zero is quite simple. It is generally based on a zero of 200 yards, and uses the rule "2 high at 22." The process is simple-- you zero your gun for a particular round, velocity and temperature to be 2 inches high at 22 yards, and it is **"close" to a 200 yard zero for a vast number of shells.**

How does this work? First, there is a near zero (PBR) and a far zero (the two places where the bullet path or ballistic curve crosses the line of sight, and you actually hit what you're looking at). For 100 yards (far zero), this is often between 20 and 50 yards (near zero). Now, for most shells, being between 1.5 and 2.5 inches HIGH at 100 yards, means you're zeroed at 200 yards. The old timers who came up with this (African guides and PH's), knew that the 375 H and H was about 2.5 (top end of the scale) high at 100 for a 2,500 FPS 300 grain bullet at 75 degrees, and came up with this "fits most cases" rule for zeroing.

Notice that this rule holds a lot of variables constant, or in other words, assumes a lot, or takes a lot for granted. Here are some of those constants that reduced loaders will think about:

1. The bullet weight and velocity are fixed. This means you know your bullet "goes to sleep" (stabilizes) normally, your SD's are very low, your pressures are uniform, smooth and repeatable, your bullet is not out of round, and the "jump" from chamber to lands for the bullet is no greater than .002, and not too much less either. In other words, NOT a factory round!! Nobody is a PH or guide for anything dangerous with factory rounds anyway. This also assumes that your powder gives consistent SD's and that you've taken temperature into account for that powder.

There are powders like **Varget that are much less temperature sensitive and powders like Red Dot that are very sensitive, so not everything is really ever "constant" with powder, and hand loaders have a real edge here in choosing their powders.** Later, we'll see the major shapes of powder, and realize that with factory ammo, the choice is not to optimize accuracy, but to choose the closest powder that meters very cleanly and hits 87% case capacity-- for mass production. The answer to this problem of course is to **ROLL YOUR OWN!**

2. Each rule of thumb only works for a single caliber.

Once you've got your "hunter's zero" at 200, and held everything else constant, you can then check your software, ballistic chart, or cheat card and start to vary other important things, like **temperature, altitude, crosswind, motion of the target, shooting uphill or downhill, etc.** You can now see why we don't want to be varying things like bullet weight at this stage, although this example (375 H and H) is unique in all the world and all the calibers in hitting very close, regardless of bullet weight.

Sticking with that example, you can have a simple scope card that shows differences up and down and left and right from 200, and you're good to go. If you have 1/4 MOA clicks with your 375 300 grainer, where you're **HIGH BEFORE ZERO (less than 200 yards), you'll be doing MINUS CLICKS (down), and where you're LOW AFTER ZERO (past 200 yards), you'll be setting positive clicks (up).** In this case, at 70 to 75 degrees, 9 clicks down at 50 yards where the bullet is 1/2" high, 10 clicks down at 100 where the bullet is 2.5" high, and 5 clicks down at 150 yards where the bullet is 2 inches high.

After the 200 yard zero, your card or software or chart or scope will indicate **6 clicks up at 250 yards where the bullet is 4 inches low**, 28 clicks up where your bullet is 30 inches low at 400 yards, and 46 clicks up where your bullet drops over 60 inches at 500 yards!

Shotgunners know the **speed of each type of bird**, and how far to lead them based on distance. Hunters know that at 200 yards, you **lead a running Antelope by 15 feet, an Elk by 10 feet, and a Deer by 7 feet**, with the 375 in 300 grain. They also know you need a 2 inch crosswind correction at 200 yards with a 5 mph wind, and a 4 inch correction with a 10 mph wind. You can get these figures from your ballistic software, ballistic cards, bullet manufacturers, scope makers, and/or ballistic manuals. A few load guides still publish ballistic charts also, and many are available online (see the resources chapter below).

NOTE: Many reduced loads use round balls, and the resources section also lists software that includes the "GS" algorithm for ballistic curves in spherical projectiles.

Terminal Ballistics in Reduced Loads

If you're hunting a water buffalo, and it charges you, you want your bullet to go directly into the brain or heart, penetrate thick skin, expand to maximum width, retain maximum energy, and retain maximum weight.

The factors that help a bullet fly through the air with less drag, like a sharp point and very high velocity, aren't necessarily the same factors that keep the bullet together, penetrate most deeply, or avoid brush deflection. This gets us to the number 1 principle in all of life: tradeoffs. All that kinetic energy and pointiness can go into breaking the bullet into shrapnel when it hits tough skin.

If, on the other hand, your target was a human terrorist, the high energy, pointed distance round might be ideal, assuming no or light body armor.

Terminal ballistics in gallery loads depends on objectives. Are you punching holes in paper? Wadcutters and round balls work as good as anything. Killing varmints? Grease or sintered powder filled hollow points can make a hole the size of a quarter with only 5 grains of powder in some combinations. Hunting? Projectile design then starts to become a bunch of tradeoffs between pressure, velocity, delivered energy, expansion, brush deflection, and of most importance in reduced loads-- range.

We're obviously not designing reduced loads for 300 yard shots, except perhaps in 50 BMG. That suggests that two candidates for projectile designs are less good (steel and copper) and one is much better (lead) due to softness. Easier on the barrel too, if correctly gas checked and lubricated!

Law enforcement? Frag and sintered rounds won't penetrate drywall or ricochet dangerously, and there are numerous dye filled practice projectiles we can make, from a whole variety of plastic casting dies. The pictures in Chapter 13 also show a number of creative explosive and other "reduced" loads that pack their punch downrange, not with noise and recoil-- genuinely silent AND deadly.

25 to 50 yards is the general limit for ultra reduced loads, and 100 yards for coyote/ deer would be inhumane except

in moderately or lightly reduced loads. Don't forget that your trade off in silence and low recoil is less energy at the finish line! This is only something to be aware of, not worry about. More on these topics in the technical chapter.

Also see Chapter 13 for examples of a few exotics, such as reduced loads with multi stage rocket pyrolants and HE tips, **for a real BANG at the end of the trajectory, after a whisper to start it off on its journey downrange!**

A Note on Rifling

We've covered case fit to chamber, and some projectile and load elements of accuracy, as well as sighting in and external ballistics, but let's not forget rifling! Many manufacturers create an "average" rifling for all loads, just as they do with chamber sizes, and the gallery loader can beat this game by designing loads for ideal rifling. The mini-30, for example, is considered a highly inaccurate "ammo dumper" in 7.62 x 39, but we guarantee it can be a tack driver with the right diameter load and the right projectile for its unique rifling. You MUST hand load 7.62 if you want true sniper accuracy, as the available ammo is all over the board in uniformity, or lack thereof.

So, what is "ideal" rifling? Most engineers default in this question to the time proven (1879) Greenhill formula, with some minor modern modifications (Miller Twist and McGyro rules). Greenhill's rule of thumb is:

$$Twist = \frac{CD^2}{L} \times \sqrt{\frac{SG}{10.9}}$$

In this formula, C is a constant (150 under 2,800 FPS, 180 over), D is your bullet's diameter in inches, and L is your bullet's length in inches (BULLET, NOT cartridge). The whole right side of the formula can be ignored, since SG is the specific gravity of your bullet, which for all lead bullets is 10.9, canceling out that side of the equation.

We all know that too low a twist causes keyholing and wild inaccuracy, and too high ruins barrels, causes sideward flight at the bottom of the trajectory, and destroys bullets. More to the point of our terminal ballistics discussion, excessive twist also robs the projectile of kinetic energy, or "stopping power" at the payload end. Notice that this twist formula does NOT take BC, or shape/ drag into consideration-- another and different factor in your load designs.

A quick mental trick to remember this formula, is to think of it as **taking a "percent" of 150 (the constant).** If your bullet diameter/length ratio is 10, 10% of 150 would be a 1 in 15 twist. This is the way we "work backwards" to design our bullets, or if we're gunsmiths, rifle our barrels.

To expand the above example, let's say we have a 50 BMG (more correctly 50 caliber) Hornady A-Max 750 grain with a BC of 1.05. What should the ideal twist be?

The diameter (.510) squared (multiplied by itself) is .2601. Divided by the length (2.555 inches), gives about .10 (or 10%). 10 percent of 150 is about a 1 in 15 twist, indeed, the twist on many 50's today. If you actually run the numbers to decimals, it comes out to 1 in 15.27-- so cast your bullets to be closer to your actual twist! But remember, if you're dealing with velocities (remember your charge) above 2,800, the constant is 180, and you're talking a 1/18 twist instead! For a "short" 50 cal cast slug only 1.5" long, the twist jumps way up to 1 in 25 (very slow twist) also-- so you can see the two factors here that raise or reduce required optimal twist.

Of course if you have a .308 engaging a .311 barrel (at the grooves of course), it's another whole problem than twist, unless you have an obturating (hollow) base to expand and meet those grooves. Length also matters not only in twist rate, but actually engaging the rifling in the first place out of the chamber, both with rifles directly, and to and through the forcing cone of your revolver.

What's the ideal twist for a 150 grain 30 caliber .308 vs. a 303 British (7.62 Russian) 125 grainer with a .311

diameter? RIGHT, grains, BC, etc. don't matter, you need dimensions!

Here are a few dimensions: the 303 British 125 grainer is .311/.905, the .308 (30 cal) is .308/1.209. For comparison, a military 93 to 108 grain 7.62 is .310/1.055.

Answers (all assuming under 2800 fps):

303 British/ "heavy" 7.62-- **16.03**

.308-- **11.77**

7.62 Military-- **13.66**

See the problem? Weight is NOT in the formula, but of course is contained in the design! **Imagine the "one size fits all" problem of barrel makers, who now will take an "average" and give you a twist of something like 1 in 10 (or 1 in 7 for an AK), reasoning that average ammo works with a twist of 1/13 or "faster."** You can either buy 20 guns, or design/ purchase your bullets to more carefully stabilize to more exacting twists!

Only loaders have this option. The mini 30 is 1/10, the 14 is 1/9-- shocking to the AK and AR shooters used to 1/7, yes? (Notice that lower numbers mean FASTER twists BTW-- 1/7 is quite a bit faster (spin imparted) than 1/13. Most armchair ballistic folks say the heavier the projectile, the more (faster/lower number) twist/spin you need to stabilize it-- but the idea of "there is no such thing as too fast" fails to recognize that **over-stabilizing keeps the bullet aligned with the bore rather than trajectory line, and at the end of the parabola, the bullet flies partially sideways, rapidly eroding accuracy**. See the RCBS Load depiction of Greenhill in the chart on the GalleryLoading website, which translates Greenhill into weight vs. twist.

What has this got to do with reduced loads? Simple, as gallery loaders, we're often dealing with shorter barrels and much lower MVs to begin with! We can get an "edge" on the inaccuracy and lower stabilization that this might cause, by "closing in" on other factors like case sizing and bullet size and length (and hence weight), to make up for the accuracy differences. If you need a really fast twist to stabilize, and don't have the time to do so in your T-C (due to short barrel length), you can design a lighter/smaller/"fatter for its length" projectile that maximizes the powder you do use, and "slows down" the twist requirement of the barrel for stabilization. IT WORKS!

More to the point, you can also cast balls that engage rifling in various ways, but closely fitting to grooves in your casting can create a surprisingly stable projectile. Cast balls can make a single hole at 50 yards with many different rifling speeds-- a close fit is crucial to make the most out of the reduced load you're using, without creating a pipe bomb with TOO tight a fit.

The "ball" logic works this way: Original minie and musket loads had VERY slow twists (eg: progressive or gain twist starting at 1/60 or 1/72 and speeding up to all of 1/48 at the muzzle). With the "x and faster" logic, up to a certain point (about 1/10 in our experience) you can't go wrong with round ball loads in most modern rifling. About .001 or .002 overcast below (above!) 1/10 (eg: 1/7), seems to work well in our tests, which is generally true of many cast loads. Remember, you can't "overspeed" a ball, because it has no pointed end to follow the wrong trajectory, and depending on relative size and channeling, it also may not engage the rifling at all. **A .4 sized average #2 lead ball will weigh about 100 grains for reference-- relatively light -- and another reason for the slower twists.**

A Note on Headspace

Headspace is the distance measured from the part of the chamber that stops forward motion of the cartridge, to the face of the bolt (with the action closed). Used as a verb, headspace refers to the PROCESS by which the correct positioning is achieved. Different cartridges have their datum lines in different positions in relation to the cartridge. For example, 5.56 NATO or .308 Win headspaces off the shoulder of the cartridge, whereas 303 British or .22 LR headspaces off the rim of the cartridge. If the cartridge is rimmed, the headspace is only the small space from the bolt face to the edge of the rim itself. If not rimmed, the "datum line" is the point on the cartridge's shoulder where SAAMI has determined the case rests, just in front of where the shoulder begins, looking from the rimless end. In that case, the "headspace" is the much longer length from the bolt face (rimless head) all the way to the shoulder datum.

The term "headspace" can be confusing, because there is really only a "head" in a rimmed cartridge (the rim at the head of the cartridge). Even so, the longer (nearly full case) measurement to the shoulder is still called headspace. In a rimless cartridge with no shoulder, the shell headspaces from the bolt face, to (at front) the protruding case mouth (not crimped-- e.g. 45 ACP), which contacts a shoulder/notch milled into the front end of the chamber to stop the round (otherwise, with no rim, the cartridge would simply slip through the chamber).

For a belted cartridge like the 375 H and H, the case headspaces off the belt and a little notch/shoulder at the rear of the chamber, creating a "rim" like short headspace measurement. The space is measured between the bolt face (back of cartridge) to the forward edge of the belt, where it engages the notch/shoulder in the chamber. Note that this notch is at the BACK of the chamber, whereas the non belted rimless notch is at the FRONT of the case mouth, right at the bullet, which of course is not crimped so it can engage the shoulder.

If the headspace is too short, ammunition that is in spec may not chamber correctly, or the bolt can become stuck, or not lock up. If forced, the bullet can be pushed into the case to the extent that excessive pressure also ruptures the case. If headspace is too large, the cartridge case may rupture (due to case stretching within the headspace), possibly damaging the firearm and possibly injuring the shooter. If excessive headspace results in case separation or rupture, the escaping gas can actually propel the bolt into the shooter's face or chest, killing him, as has happened with some 50 BMG's that do not have lugs securing the bolt (also called slapback). In addition to having a bolt shot into your brain, a vibratory shock from a relatively loose bolt combined with excessive headspace (as in milsurp rifles) can also shatter the receiver into grenade-like shrapnel pieces that then blow your face off, even if the bolt is stopped by the receiver (the pieces come from the kinetic force of the bolt slamming into the brittle receiver).

When the cartridge fires, the primer backs a little out of its flash hole, then is forced back in by the tight bolt, and excessive headspace also can stop this back off/ return action, exploding the flash hole and head.

Headspace is measured with a set of two headspace gauges; a "go" gauge, and a "no-go" gauge (see pictures in Ch. 13). Modern headspace gauges resemble the cartridges for the chambers they are designed to headspace, and are typically made of heat-treated tool steel. Both a "go" and a "no-go" gauge are required to headspace a firearm properly.

Headspace gauges are typically used by inserting the gauge into the firearm chamber. The bolt should close and lock on a "go" gauge, and not close on a "no-go" gauge, indicating that the chamber headspace of a firearm is within safe minimum and maximum dimensions, respectively. The force that is applied to the bolt on a bolt-action firearm when making these assessments should only be at normal levels of force; otherwise, an incorrect assessment of headspace may result if the bolt is forced into a position with excessive pressure.

For current or former military calibers, a "field" gauge can also be used. The "field" gauge is designed to take the place of the "no-go" gauge in military firearms, and functions in the same way. Military firearms are designed to withstand higher pressures. As such, a greater tolerance in the firearm's headspace is acceptable, and the "field" gauge takes into account this greater tolerance. "Field" gauges should be used only on military firearms, and not on commercial firearms. Headspacing a commercial firearm with a "field" gauge can create an unsafe condition.

As the "field" gauge takes the place of a "no-go" gauge, any military-surplus gun that locks on a "field" gauge is unsafe to fire, and should be checked by a trained gunsmith.

Guns that fail to lock on the "go" gauge may simply need cleaning, especially at the bolt face, as build-up may occur on this surface and this buildup can cause problems in chambering a round without stressing the brass.

Headspace gauges are designed to indicate simply whether a firearm's chamber is in tolerance. There can be other types of gauges that measure exactly how over or undersize a chamber might be. For example, a bullet engages the lands at the bullet's ogive, not the point, and a slight "jump" to the lands is important for pressure (usually no smaller than .002). Direct contact with lands increases both pressure and accuracy, but can result in serious pressure problems in some loads. OAL measurement gages combine with bullet ogive caps on your micrometer to get exact OAL measurements for that particular round, in order to exactly design your round for your optimal land distance.

Remember, the headspace gauge itself is not enough-- the chamber, gauge and cartridge must all be designed to the same spec! The two biggest problems with headspace in the U.S. are surplus ammo and surplus firearms.

Since the case and head separate just in front of the head, a ring and/or bulge in that area on a fired case can be an early indication of headspace problems, as can primers that protrude after firing (also an indicator of too low pressure, as well as a bolt face that is not reseating the primer due to headspace or bolt issues). Along those same lines, there are gauges that check for headspace problems by caliper measuring stretch at the head, and little wires that you can use to feel the inside of the case for a small groove that indicates stretching near the head, prior to catastrophic head separation. Frequent misfires are another indicator of possible headspace problems. "Play" in excess of .004 to .006 is considered excessive in most rifles.

Remember, regardless of chamber gauges, headspace is the RELATIONSHIP between the cartridge and the chamber, and you can have headspace problems due to improper sizing of cartridges, even without a headspace problem in the chamber itself! The go no/go gauges have nothing to do with this aspect of the problem.

SAAMI can also be a problem due to too wide a range. Take a .004/.013 example. In this example, SAAMI maximum headspace/minimum cartridge allowance produces so much clearance, .013 inch, that it will reduce case life. This much headspace would probably be found only on a used rifle. The SAAMI minimum headspace/maximum case condition produces .004 inch of interference, which could make semi-automatic operation unreliable.

What has headspace got to do with reduced loads?

1. Reduced loaders may have more of a tendency to full length size their brass, which can make it shorter or longer (out of spec), creating excessive or insufficient headspace, due not to the chamber, but the ammo. Fix: neck size, and check optimal lengths to specifically "design" each load to a particular rifle chamber, bolt AND lands. Handloaders have options factory folks don't, in exactly fitting their ammo to a particular chamber. See the pictures for some important tools Smith Tactical uses to accomplish these custom fits.

2. Reduced loads themselves can cause insufficient primer return and protruding primers/ bulges.

3. Many reduced loaders use older weapons, curios, period pieces, re-enactment guns, etc. These can have OTHER features that increase pressure, and have completely non-SAAMI chamber specs, or use ammo with widely varying specs. Give your gun a thorough run through before designing your cartridges. Cerrosafe ("slug") your chamber

and barrel for exact specs and measurements. Use calipers, chamber gauges, go/no-gos, ogive/land measurements, etc. **before firing**. (BTW, "slugging" a barrel can both mean taking a measurement with molten metal, AND pushing a lead bullet through the barrel to "feel" for problems, wear, etc. The term is used in more than one sense). If you're not a gunsmith yet, consult one before trying out a "new old" gun!

4. Headspace pressure problems can compound "start-stop" features of supersonic deflagration, changing an annoying problem into a deadly one. If you use a non-published reduced load which gives you bullet jam and start/ stop ("pipe bomb") problems, and also have primer and head bulge/separation issues going on at the other end, the combination can be catastrophic. Most problems ARE additive, often very quickly, and exponentially. A brittle receiver on an old gun, a gun with weak closure like the T-C, etc. when combined with excessive pressure, detonation, AND a headspace issue, can mean the difference between the explosive event destroying a gun and destroying a face.

5. Indicators, such as shiny rings near the head, primer indicators, bulges, etc., can indicate BOTH reduced load problems and headspace, bolt, powder and/ or chamber problems. The key to safety, and research and design, is **isolating variables**. We DON'T WANT both reduced load and headspace indicators going on at the same time! Make sure **everything else in the system/ cascade/ harmonics is stable** before testing your reduced loads!

Standard Deviations (SD)

When you have low relative velocities, as seen in reduced loads, a high variance in FPS (Foot Per Second Velocity) (lack of uniformity) can be a pretty high percent, because the denominator is smaller. In general loading, other than Bench Rest Shooting (BR), or sniper applications, SD doesn't mean all that much-- groups and pressure curves are what matter. But, as we'll also see in the technical/ engineering chapter, in reduced loads it **also is a measure**

of risk, just like in many other fields, such as research and portfolio management.

In the old pre-calculator days, and in some places still today, load developers and loaders used the "extreme spread" (ES or XS) to measure variance, or lack of uniformity. Half of the ES was supposedly an approximation of what we now call SD. The ES is simply your highest speed in a group minus the lowest.

The formula for Standard Deviation is:

$$s=\sqrt{\frac{\Sigma(X-M)^2}{n-1}}$$

This looks intimidating, but is really just the "sum of squared averages" (or means). The "s" means standard deviation, the "big E" means add the squared differences together (each try, X, minus the average, or M, the mean), divide them by the number of "tries" minus one, and take the square root of the whole thing.

We know, **all ballistic software programs, and most chrony's now just give these to you with no effort**, but it is good to see how the measurement really works, since very high SD's (over 40+), can mean a very risky load when we're dealing with reduced loads. Think of it as a percent. Manufacturers see 25 to 35 (now that everyone has a chrony, just 10 years ago factory ammo with SD's in the 100's were common!) as a good SD for velocities around 2,000 FPS. Just looking at a linear percent, this is 1.75 percent. The same percent-calculated SD for a load shooting at 680 FPS would be 11.9%!

Granted, SD's are non linear, and hence weighted, but you see the point-- the number is still more sensitive if you try to apply a "generally good" SD (no such thing) to all cases. Even in reduced loads, SD's below 25 are considered quite stable and uniform, but our goal should be single or low double digits for really stable loads outside of manufactured ammo with very deeply reduced loads. Even an SD of 35 is low risk and technically safe in a reduced load, and won't be keyholing at close range, or shooting wildly off from shot to shot, assuming pressure curves are also tame, but we **bold face loads under 30** in our charts as passing a basic gut check of better than industry standards for their own uniformity in a vast array of firearms. **Also don't forget: in a reduced load, even a fraction of a grain can have a MUCH bigger impact on SD than in max loads**, as a function of the percent it contributes to overall burn uniformity and pressure. The **difference between a slight metering error in starting loads and reduced loads can mean a destroyed gun**-- always HAND WEIGH your gallery load charges!

Many will argue that for the short range work involved in reduced loading, SD's don't really come into play-- and they are right from the standpoint that a one inch group at 1,000 yards vs. a three incher at 50 feet are irrelevant if you don't care about 1,000 yards, and only care about absolute groups. However, the other factor-- **how uniformly your load is operating (group to group, not just in a group), is an important indicator of the safety of your burn-- and is REFLECTED in the SD, as well as pressure curves, as a measure of both uniformity and risk. The Engineering Chapter gives more information on actual pressure testing instrumentation, which is a major step beyond SD in ballistic lab operations, and which is now available in PC versions to load developers with home shops or gunsmiths/reloaders moving into ballistic labs.**

Sure, the tighter your SDs, the tighter your groups, but why care about anything BUT groups? Because small SD's (single digits, low doubles) **over many trials**, can mean a very safe and stable powder burn and lower likelihood of problems, and when also compared to no "second spike" pressure indication, can avoid everything from detonation to stuck bolts to chamber damage or barrel warping. **If you try 5 shots and get an SD of 15, after 80 to 100 you may find that the actual SD for that load is 25**-- sampling error is rampant with loaders who want to either move on quickly or save money on ammo. 25 is still respectable, but if the jump is from 30 to 80, you could be in trouble-- 80 is a 10 percent variation in an 800 FPS load, and could also be an indicator of a second pressure spike caused by incomplete burn, secondary burn, and/or an obturation detonation event.

There is no single barrel durability spec, and no single "safe" pressure limit on ANY gun-- each needs to be evaluated individually according to the manufacturer's specs and condition of the barrel. SAAMI will stay very "safe" in telling us that a FACTORY NEW 30-06 will get into trouble beyond 50,000, when actual barrel damage starts at 80,000, and deadly destruction occurs at over 100,000 PSI. BUT these numbers, and the pressures actually produced, also can come from damage or wear on the barrel-- **as in a very worn barrel having much lower destruction thresholds, AND much higher pressure DUE TO the wear.** We've seen catastrophic failure in a 30-06 that developed 166,000 PSI with a load rated at 40,000, due to barrel wear-- the load isn't the whole story. Spikes and irregularities also result from how close to ideal land contact each round makes, and slight (thousandths) differences can result in lower and higher pressures as the bullet "jumps" or "overshoots" the lands as it's freed from the neck. **The bottom line: NEVER exceed a published load, PERIOD!**

Of course there are many other factors that can affect shot to shot consistency and the goodness of fit (overlap) of your pressure curves. **One that is almost never spoken of is heat.** Give each shot a good rest before the next in order to cool the barrel, so barrel temperature does not become a co-variate problem in measuring your load performance! Remember, holding as many variables constant as possible is required to isolate any other variable in load development research and testing.

Factory ammo, from single batches, varies between 25 and 35 SD, in thousands of loads tested (AFTER the mid-1990s). Hornady consistently hits in the low 20's for many of their factory rounds, and with ammo matched to a gun by a competent reloader, Hornady components consistently drop to below 7! We're not particular Hornady "fans," but they do make a number of quality products. Match shooters and ammo **expect SD's** to be single digit, and pressure curves to literally overlap. **With gallery/ reduced loads, we have to "cut our loads some slack,"** and be pleased with SD's that don't go much over 35, and very pleased with those that are under 25 to 30 over many trials. Under 30 is far safer, in general terms, than over 40! Notice that without a LOT of shots, the two numbers are actually within the margin of error of the test itself in many test designs. There are ways to reduce trials and use "student T test" and pressure curves instead of SD's, holding loads to a minimum standard, covered in the engineering chapter. ***Needless to say, if a published recipe showed indications of secondary or high pressure spikes, regardless of SD, we didn't include it here, even if it is an "accepted" load from a reputable manual! Those who simply publish the loads of others without testing them aren't adding much value other than convenience. That convenience itself can be dangerous if you don't take responsibility for checking your own loads with SD's and pressure curves!***

Let's say you shoot a 12 gauge subsonic slug load **(2.75 Hull, 209 primer, 10 grains of Titewad, 1/4" felt wad, overshot card, 8,000 psi, trying for 900 FPS, folded crimp)** five times. We shoot a minimum of 80 trials for each SD, but this is just an example. The speeds you get are chronographed at 986,992,980,976 and 990 (actual numbers from one of our trials). Your "N" (number of tries or shots) is 5. The average (or mean) is the 5 numbers divided by 5, or 984.8.

Your XS/ES (Extreme Spread) is 992-980, or 12, so our gut says we're at around 6 for SD. (Don't rely on this rule of thumb, it is easy to get an ES of 24 with an SD of 9, etc.). Taking the average (984.8) into the SD formula gives (986-984.8)^2 + (992- 984.8)^2 + (980-984.8)^2 + (976-984.8)^2 + (990-984.8)^2 or 1.2/7.2/-4.8 (squaring will eliminate the minus)/8.8/5.2, or squared, 1.44+ 51.84 +23.04 +77.44 +27.04 (Total= 180.8), divided by the number of trials minus 1 (n-1), or 4, which equals 45.2. Take the square root, and the SD comes out to **6.72**. **Very nice, but way too low a sample size! At least gets marked as a "promising" load for more trials and potential instrumented pressure curve testing. By the way, *Red Dot (we won't publish how many grains) vs. Titewad in this test, gives SD's over 50!* As similar as they may seem, Titewad (in this case) is far and away the safety and performance winner! MORAL: DON'T EVER generalize load to load! Red Dot smokes Titewad in other tests and combinations! If you've seen one, you've ONLY SEEN ONE.**

Other factors that can cause discrepancies in SD stats in addition to incorrect sample sizes include chronys positioned differently or incorrectly to standard (SAAMI gives the middle of the chrony at 10 feet as a standard, then publishes speeds for 1911's taken at 25!). All our tests are always at the 10 foot distance, unless requested differently by a customer or armorer. Some amateur testers think "muzzle velocity" needs to be at the muzzle, and place the chrony too close. (Blowing a chrony away can also make one a little skittish about placing it too far from the bench)!

CHAPTER 5. CHOOSING A NICHE

Whether you're getting into gallery loading for fun, as a hobby, to make some side cash, or to start a serious new career or business, we encourage you to specialize!

The old Boston Consulting Group "smiley face" of market share vs. ROI goes:

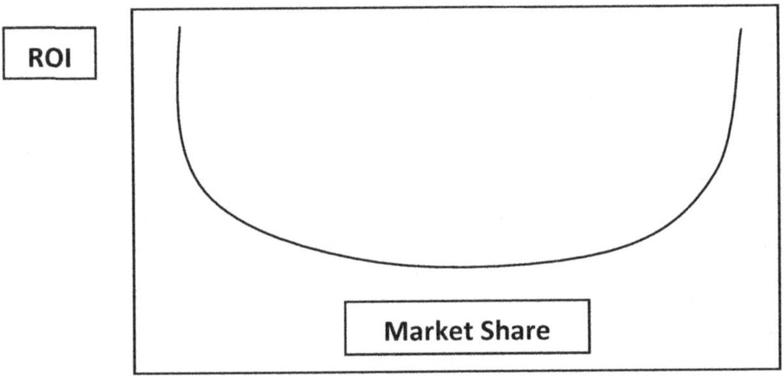

This simply says that, **with either very LARGE market share (big company monopolies), and with SMALL market share (NICHE marketers), you get GREAT ROI's, and successful businesses.** In the middle, are all the price takers trying to be everything to everyone, middle sized, cutthroat competition, and losing money by the truckload.

In everyday language, it says, if you want to sell ice cream, and you sell every flavor, you'll be invisible among all the buyers and sellers, but if you specialize in pistachio, and TARGET your exclusively best niche product to your exclusively best niche audience (pistachio lovers), you can be the BIG fish in the smaller pond, and OWN your market.

If you're a gunsmith or reloader, specializing in gallery/reduced loads can be a real win!

Even then, you can further specialize by type, audience, gun, load, projectile, customer, etc. Why not become the best custom .410 single shot seller on the planet? Even more specialized-- of buck and slug loads?! The list is endless-- cowboy action shooting, fast draw, accuracy and sniping, police and military, training, home defense, varmint control, animal control, hunting, "city" loads... etc.

We strongly recommend that you join AAGSR (American Association of Gunsmiths and Reloaders-- AAGSR.org) for not only many ideas on continued success, but also solid leads and help in launching and sustaining your niche!

If you're just loading for fun and as a hobby-- specializing still makes a lot of sense, as there are only so many hours in the day. The "questionnaire" goes something like: -- Do you like rifles, handguns or shotguns? -- Are you into hunting or home defense? -- Do you like long range rifle or short range handgun best? -- How old are you? -- Are you interested in working with youth? -- Where do you live? -- What is the LOCAL market like? -- What equipment, firearms, and components do you have, and what can you get? -- ETC.!

Many folks have made really good "hobby" income specializing in new technology support as new guns are launched-- like offering new products surrounding the Taurus Judge-- loads, holsters, ammo, accessories... etc. **Heck, a revolver that shoots a shotgun load is right down the middle of a gallery loader's sweet spot!!!**

In addition to loads, you can also specialize in selling tools, equipment, helpful aids, accessories, and even kits. There is one enterprising smart young guy in Canada (Brian Hammond) who has created the "game getter" (we think he may have gotten the idea from one of our previous editions or posts about ball loads-- more power to him, he's a great, ethical, smart guy). The "game getter" is an entrepreneurial empty rifle case drilled off center. You swage a shotgun pellet into the mouth opening, and use a **nail gun .22 blank** (brown or green, brown best!!) inserted into the offset "primer" hole, and can get accurate "ball" loads up to 900 FPS!! Why is the hole offset? Because the set up is striking the rim of a nail gun blank with the center fire firing pin of a rifle! See the pictures for more details.

Chapter 6. Loads and Techniques

This chapter lists a wide range of time proven, manufacturer tested gallery, reduced, and sub-sonic loads, filtered by decades of Smith Tactical SD and pressure testing and research. Not all loads are under 1130 FPS, but all are far under the starting loads in other more general manuals. Every load in this book was taken from a published, out of print, or public domain SAAMI TESTED source, including testing labs, bullet and powder manufacturers, and specialty equipment, component and ammunition makers and testers.

The unique to this manual SD's were exclusively researched and pressure tested at Smith Tactical for this volume, as well as for contracted customers. SD's are solely for the purpose of getting a start on which loads appear uniform in testing, **to save you time and money in avoiding loads that are dead ends.** There is NOT a perfect direct correlation between SD and group size, although they are usually closely correlated. Some loads group a little better with higher SD's, but in many cases that is due to not enough trials (too low a sample size).

Every SD was augmented with pressure curve trace tests, and **any load** showing pressure problems, **low uniformity (disparate pressure curve overlap), or secondary spikes, flat or unexpected peaks, etc. was eliminated, even if it is a known, reputable load.** Note that in "population sized" samples like the large number of trials we report here, SD is based on n instead of n-1 in the calculation denominator. The better ballistics programs-- such as RSI Shooting Lab-- will convert your Chrony readings to the population SD automatically if you, like we, believe in large samples North of 60-80. We also use the much more modern algorithm models (such as RS for round projectiles), since manufacturers don't generally do much testing for gallery loaders as a specific audience and application.

We exhaustively researched EVERY PUBLISHED SOURCE of reduced loads, and **if you have a pet or favorite load that is not listed here, it likely did not meet our minimum SD and pressure trace uniformity quality standards** for inclusion. It might be a perfectly good load for its purpose, but many supposed "tested" loads did NOT stand up after thousands of trials over many years, and appeared to be based on too small a sample when they were reported with acceptable SDs and rarely pressure traces in other publications. In fact, Smith is the largest tester of reduced loads, by a factor of hundreds over any other ballistic lab, and many previous published loads are based on math, not real pressure testing, which is dangerous and unfortunate.

Statisticians will argue that if SD's are worse with higher samples, it might be due to a truer picture, or other variables being different over many rounds. Either way, ***it's not worth wasting your time or health on loads that keyhole or present potentially dangerous pressure curve burn signals.*** **Since we augmented each SD test over many years with pressure testing, the loads we have left out are not worth your time, due either to accuracy problems, or dangerous variability. If you think we missed one, email us, and we'll either test it, check our database, or let you know why it was problematic. See the pictures for some of the sophisticated tests we employ for EACH load we research or design.**

If a lot of the loads here group between 20 and 30 FPS SD, it's not because they are all like that in every batch (many were as low as 0-6 in handloads, and as high as well over 100, right from the factory!), it's **because that band of SD's are the ones that we chose to include, due to proven**

uniformity over many thousands of rounds, over two decades of testing. For many reasons, we're not including reduced loads that turned out poorly, **even if published in leading manuals**. We don't want a respected manufacturer looking at one of their published loads and saying... say WHAT??

If you're a manufacturer, and would **like a load included that we skipped, or that's new, email us at Mail@ReducedLoads.com**, and we'll work with you on a new test, or share more sophisticated pressure curves and AUC's with you. If a load consistently hit over 50, or had widely varying pressure curve profiles, why publish it when we don't trust it ourselves, even if it's published by a leading manufacturer? If you can't afford expensive oscilloscopes and other pressure instrumentation, **Smith will test 11 rounds of your favorite load for $50 bucks, just email Mail@ReducedLoads.com with the subject: Pressure Test Request.**

To be very honest, **before the advent of chronys being used by many more reloaders, especially by numerous home reloaders, bullet manufacturers were all over the map.** Their loads regularly had SDs over 100+! With chronys "for all of us" becoming the norm rather than the exception, manufacturers were forced to raise their quality and uniformity exponentially. Sad, but true, as most industry insiders already know.

With the coming advent of ultrasound, xray and other shape measurement gages, **another revolution is about to take place.** Perhaps 1 out of 10 bullets sold by even the best manufacturers are uniformly symmetrical enough, inside and out, to be considered match or sniper quality. Less than half are even precision hunting concentric! Just like when chronys took over, when symmetrical uniformity measurements become the norm, manufacturers will be "busted again," and forced to greatly tighten their uniformity standards. See the pictures in the last chapter for more information on doing your own run-out testing.

By "fixing" other variables like powder, bullet quality, uniformity, chamber and land fit, and loading technique, whisper loads can be FAR MORE accurate than the average factory load, within their range limits. This includes quality primers, the right, carefully measured powder, fire forming the case to the chamber, completely symmetrical bullets, and length that correctly spaces to lands, etc. *See the "Madigan Pyramid" from Smith Tactical in the Engineering chapter for more variables like human factors, age of the barrel, rifling, etc.*

As only one example, Alliant's 410 was looked upon with enthusiasm as an ideal new candidate for reduced loads, when it was first launched. We, however, found over many trials that it has some of the **same problems as 2400 in uniformity, if reduced too low**, even though its speed is very close to Lil Gun. The jury is still out, and it is still an outstanding choice alongside Titewad and LilGun for many 410 loads-- **some of which are uniform and "naturally" close to subsonic**, just due to the speed of the .410 bore/ shell itself. Of course the 209 primer itself isn't exactly silent, as cowboy action shooters know! We would **NOT** put 11 or 12 grains in a 30-30 shell, however, unless we were suicidal!

Of course it's not just the cartridge. Regardless of "SAAMI rating" for a particular firearm, we've got to also use a little common sense-- if we get pressure problems in a TC with their anemic little closure lock, or in a 50 BMG without positive locking lugs for the bolt-- projectiles, gun pieces and gas are going to come AT us rather than go up, down, or forward. Worn barrels, obstructions, incorrect headspace, bad primers, and many other variables can cause overpressure problems-- **good load developers**

ALWAYS pressure test their loads with PT traces, in addition to SDs. While we're on the bandwagon, has anyone else noticed how terrible primer quality has degenerated in the past decade? We hate the thought of making our own primers, but if the trend continues... There's another reduced load risk factor you won't read about anywhere else-- you heard it here first. We'll bring you ongoing updates, as Smith tests more primers, on the website. CCI is still holding their own, as of this writing.

To keep your manual up to date, we will publish new loads on GalleryLoading.com's private user pages, which you can access as a purchaser of this manual. The User ID is gallery (all small letters), and the password is 97539, at the time of this edition's printing-- **email us to stay up to date**.

In addition to the loads that follow, **there is a "technique" section at the end of this chapter with a lot of little known FUN tools and techniques out there,** that represent the best in entrepreneurial innovation, as well as suggest areas where you might wish to experiment, research, design, test and develop new techniques, tools and products in the rapidly growing area of reduced loads.

Pressure Test Results

In addition to the **only published source of reduced load SD's**, this manual also has the **only complete file of pressure tests for each reduced load**. Our traces are organized into six categories:

1. Excellent -- Pressure traces overlapped in outstanding fashion over many trials, with no signs of spikes, excess climbs, wide variation, or dangerous secondary spikes.

2. Good -- Some variation, but still well above factory ammo in results.

3. Acceptable -- Variations can affect accuracy and grouping, but showed none of the features of unexpected deviation from uniformity, SAAMI pressures, or warning signs of uncharacteristic burn.

These three above are the only published load formulas we included in this manual. The other three categories are:

4. Unacceptable -- In addition to accuracy and grouping problems, these traces showed wide variability, lack of overlap, and differing shapes from shot to shot. We do not consider these safe, accurate, or worth further testing.

5. Strong Warning Signs -- These recipes showed alarmingly high pressure spikes, very different shapes shot to shot, and nearly no overlap uniformity from shot to shot. SD's were often well over 100 for these traces as well. These trials were completed in explosion chambers, and had the potential to damage even new firearms.

6. Dangerous Results -- Very disturbing traces, causing the trial to be abandoned. These showed secondary spikes, pressures that went "off the chart," and traces that were reversed, inverted, and/or looked like completely different recipes shot to shot. Many of these were with otherwise "good" reduced load powders, but outside of their "sweet spot" due to ultra reduction in load. Others were known culprits, slow burning, or misapplied (in our opinion) to the weight of the bullet. Many were also with jacketed rather than lead bullets. VERY FEW with Dacron fell into this category, unless the load called for a very slow burning powder. NOTE: ALL the dangerous loads ARE in published manuals today, although some have now been updated with newer editions.

Email us at Mail@ReducedLoads.com if you have a specific question on one of your own favorite loads, bullets or powders that does not seem to be here, and we'll let you know, if we're able, if it dropped off the charts due to problem traces and/or SD's. **To rate a 6, the load had to have BOTH very poor SD's AND very dangerous PT's.** We are not publishing the dangerous loads due to liability reasons, as we don't want the rare Darwin award winner trying them out. Also, some of these loads have been involved in accidents where we have been called as expert witnesses, and we're bound by court order or contractual agreement not to disclose the results.

IMPORTANT NOTE and DISCLAIMER: **Just because a load performed well in our test guns DOES NOT GUARANTEE it will be safe under all conditions, in your guns, or where many other variables can influence the outcome. If you see ANY indication of your own SD's being high, or bad pressure traces, IMMEDIATELY DISCONTINUE your development work on that load! ALL LOADS listed below are considered "safe" by the manufacturers who published them, and are all tested by them to SAAMI specs. The fact that we found discrepancies even with listed, published recipes should be an indicator that even published loads are not always safe, and also depend on the gun, the loading technique and quality, headspace, case length, crimp, primer, and many other variables we or the manufacturers cannot control. Please read the important safety information and disclaimer at the beginning of this guide.**

Also remember that FACTORY AMMO has, in the past, had far worse SD's than is the current practice. **Private loaders also have the ability to reduce SD's to single digits, which factory ammo rarely achieves even with maximum loads of the "best" (most expensive) match ammo. We've seen SD's over 40 very frequently in some of this supposedly best factory ammo! According to our BR shooters, this situation has been getting better with ubiquitous chronys showing up across the land-- but it's sad that those had to be the "necessity" factor that has**

driven the manufacturers to closer die tolerances and GMP.

Example Problem Pressure Profiles

The following notes are from our own pressure profiling tests, and **ONLY apply to reduced loads.** If a particular powder/ caliber combination indicated risk in our tests, it does NOT mean that the powder isn't perfectly fine in min or max loads as published in reloading guides and manuals from the powder and bullet manufacturers. At this printing, our database has over 200 recipes that show problematic pressure traces. Email us at Mail@ReducedLoads.com for info on your own favorite load, or to request a test of a load.

Instead of publishing an extensive list of what you SHOULDN'T try, below are a few notes on combinations that didn't work well-- please check the listings above, or the Reduced Loads website, for those that did much better! Email us for info on your own favorites. We email readers, and post replies on the website, to keep your information up to date.

-- **IMR 3031** didn't fare particularly well in 375 H and H, except in a narrow band of cast bullets using only 45 grains (246 grains, cast similar to the one in the charts above (at 260), using **Reloder 7** in 30 grains-- the **RE 7 was still much more stable**. Higher and lower grains gave worrisome signals in RLs.

-- **4064 and 4350** are poor choices for 30-30 cast RL's compared to the old standbys of **RE-7, 3031 and 4895**, as are slow burners in .308. See the test chart section on "30 galleries" for **3 grains of Red Dot and 4 grains of Unique** recipes for very small cast loads (90 grains) in a variety of 30 cals. Although you'll see **4198** in the chart for reduced cast loads, with lead, it also gets better and better as you increase velocity, and is a great cast choice with excellent pressure profiles. Naturally, it's great for .308 too!

-- In .308, **3031** performs very poorly in reduced loads, but better and better as you "load up." It has proven in our tests to be VERY density sensitive, so RL's beware. Very FINE powder for hotter loads.

-- In 30-06, **4895**, generally awesome, gave some troubling indicators in RL's with cast bullets. **4198 and 4227** were FAR more consistent in SD's and pressure profiles.

-- **4350 and 4064** have long been known to register troublesome SD's in 45-70, and this is no different in RLs. **4831** (both versions) (using a patch with cast bullets) is an old standby, but extremely dirty in RLs unless you're careful to Dacron/ patch it up, just like you'd do with black powder. We could write another whole manual just on chamber and barrel fouling with RL's, but Dacron solves a lot of those problems, as does using faster burning powders. "Rings" and other chamber, barrel and cartridge telltale signs are a different matter, pointing to pressure problems rather than just unburned powder.

Notes on Lee, and the various Internet Cast RL Formulas

Lee Precision Reloading is hated by many competitors, and loved by many reloaders, for offering low cost, high quality tools and components. Even more jealousy is directed at Richard, Robert and John Lee for their extensive work on reduced and gallery loads, yet they have contributed enormously to our science!

As you'll see in the charts, many reduced loads are based on purchasing or casting lead projectiles. If you purchase a load, you will normally get, or can request, a Brinell Hardness Number for that bullet or lead formula (BHN). Hardness can be converted to maximum bullet compression pressure for that load (and you should load 15% below that number for safety); simply multiply the BHN by 1,422 to get **max compressive pressure for that**

bullet hardness. (Notice this is BULLET pressure tolerance, NOT the pressure rating of your firearm, or the pressure generated by the load! You can have a gun rated at 50,000 PSI with a load that develops 45,000 PSI and a bullet that disintegrates at 25,000 PSI -- three different, but obviously related, measurements. Also, don't forget to check and standardize (compare apples to apples) the TYPE of pressure index-- PSI, CUP, LUP, CIP, etc.).

Most ballistic labs, including ours, have at least half a dozen state of the art quality BHN testers, varying from $900 to $2,800+ in price. **For about the price of a set of dies, you can get the Lee Cast Bullet Hardness Tester (Kit)**, which in our experience, once mastered, produces results within thousandths (insignificant differences) of the most expensive metallurgical lab equipment available for your cast recipes.

In the resources chapter, i.e. on the ReducedLoads.com website, **you'll discover links to a number of online "calculators" for ANY reduced cast load.** These generally take the Lee Modern Reloading manual "1 Grain Factor" ratings, and reduce as far as you'd like for that bullet weight and powder. You can also load Lee's own factors into Excel from page 159 of their Modern Reloading Manual (2nd Ed). Lee uses a factor of a 2.5% pressure reduction for each 1% in charge reduction (thousands of our own loads put this at 2.473% since Rich Lee published his data in 2003 -- pretty darn close!).

Although Lee gives most of this valuable info in a chapter on reduced cast loads, **he doesn't have time or space to get into the crux of the issue: reduced loads mean significantly lower velocities,** meaning significantly less accuracy beyond a certain point (mathematically, a non linear increase in SD's below a certain pressure AUC and spike shape). He also doesn't cover "detonation" (rapid deflagration/ non indexed pressure), or risks as measured by either SD's or pressure traces. The issue is powder burn speed. Slower powders can be problematic for any reduced load, jacketed or cast. Yet, from an efficiency standpoint, **slower burning means a longer duration at peak pressure**, meaning *more velocity will be generated at the same pressure with a slower than with a faster powder. The "rocket" powders, like 50BMG, also are the slowest burning.* We see this not only in our good traces, but especially in our BAD traces, where **maintaining that peak too long CAUSES or exacerbates the secondary peak** or excessive pressure combustion chain reaction/ non indexed blockage in too reduced a load.

What Lee doesn't cover is more than made up for with his Brinell tester-- a low cost way to completely eliminate one (somewhat rare) category of risk: cast bullet blockage due to disintegration or blockage under pressure. We don't know conclusively if "too hard" a cast load will turn a nice cast projectile into a blocking problem akin to a jacketed solid, but some early research shows that it will at VERY high (limit) BHNs, as in over 32. This is less of a problem than disintegration, because it is not easy to accidentally engineer or cast this hard a lead projectile (you'd need almost 50,000 PSI in a lead projectile's compression formulation strength to equal the 35 BHN of copper-- average cast lead bullets are generally half that). **In other words, a bad match between, caliber, load and bullet compression strength is not a good thing. To put it positively, *if lead rounds are forgiving in RL's, they are even more so with the correct BHN!***

We're not bemoaning the trade off, and in fact there are "creative" solutions we use with the military like 12 gauge "rocket" rounds that have a **secondary burn chamber** (see Ch. 13), AFTER the reduced load exits silently, with no recoil! But for our average weekend gallery outing, we simply find a middle ground by **eliminating worrisome**

traces and SD's, while staying carefully within BHN pressures in cast loads.

Notice that the formulas from Lee and on the web are actually virtually infinite-- you can use factors for any bullet, including any you cast yourself, and **come up with your own "reduction" recipe. *This can be dangerous if you don't include the other factors like SD and pressure uniformity-- use care, and always chrony your loads!*** However, Lee DOES ELIMINATE one potentially serious source of risk-- loading **too hot a pressure for the bullet compression strength itself** (in addition to our many discussions about chamber/ action and barrel PSI limits).

Our engineers will argue that a lead projectile disintegrating under pressure is still safer (as long as you check the barrel afterwards) than a copper projectile jammed or turned at the lands, creating a pipe bomb. However, the shape and blockage of a disintegrating projectile, the nature of the start/ stop, and effect on a non-indexed combustion (flame front) chain reaction, are completely **unpredictable round to round** with excessive pressure spikes.

DO calculate or test either your purchased or personally cast lead bullets for BHN and max pressure! You'll find surprising results, like pure lead disintegrating once in a while at not much over 10,000 PSI (round projectiles are an exception due to shape); #2 with varying amounts of antimony and water vs. air cooled at 15 to 20K; Linotype at 25K; Monotype at 30K; Heat treated (including #2) with various techniques at 40,000 PSI, etc. Of course we're talking safety and accuracy here, if you also are thinking about barrel fouling, don't forget proper lubrication, Dacron wad, and gas checks.

One flaw in the whole hardness scheme is the effect of gas checks-- a tiny copper cap can mean a lot both to fouling and integrity of the projectile. DO hardness test at the tip or ogive/ side, NOT the base, regardless of check. At the velocity of reduced loads, checks are not normally used, but they can also be important vs. pressure and obturation AFTER land engagement for accuracy, in addition to preventing fouling. See chapter 13 for some tools for length/ land engagement measurements. Always measure at the ogive, as tips vary greatly, even (or especially?) in factory ammo.

In summary, Brinell testing is fun, and will soon become second nature as another inexpensive tool in your RL kit. ***It also will greatly increase your own creativity, credibility and business opportunities.*** For example, many blacksmiths and bladesmiths know that a variety of **metal re-crystallizing procedures**, from annealing to hardening, tempering, and from water to cryo quenching, to air and oven-stepped temperature "curing," can have greatly varying effects on hardness-- and consequently **safer pressures and much greater accuracy** with reduced load cast bullets, once sorted by grains, powder and caliber.

Mix up your own formulas, plus or minus lino/wheel weights/ antimony etc., then try various cooling and hardening techniques, and test for BHN, group size, SD, and pressure! Ultimately, you'll be designing loads for maximum accuracy at the load levels you or your customers need, custom tailored for that gun and caliber/ load. More is unknown that known about this today-- a GREAT opportunity for your own development work and creative product and service offerings.

Please also see the illustrations / photos in Chapter 13 for many interesting reduced load tools, in addition to the research results below:

Caliber	Powder	P Grains	Bullet	B Grains	FPS	SD fps	Pressure
375 H&H	**Blue Dot**	17	#2 Lead	250	1350	**24**	Good
375 H&H	Green Dot	11.5	#2 Lead	250	1150	36	Acceptable
375 H&H	Red Dot	10.5	#2 Lead	250	1120	33	Acceptable
375 H&H	Herco	15	#2 Lead	250	1320	23	Good
375 H&H	**Unique**	13	#2 Lead	250	1225	**22**	Excellent
375 H&H	PB	10.5	#2 Lead	250	1100	26	Good
375 H&H	**700X**	10	#2 Lead	250	1100	**29**	Acceptable
375 H&H	SR 4756	18	#2 Lead	250	1425	33	Acceptable
375 H&H	630	20	#2 Lead	250	1400	32	Good
375 H&H	**SR 4759**	34.4	JSP	220	1925	**27**	Good
375 H&H	H 4227	35.6	JSP	225	1895	34	Good
375 H&H	**SR 4759**	26	Jspitzer	235	1485	**28**	Good
375 H&H	**SR 4759**	31.2	JSP	285	1400	**26**	Good
375 H&H	**IMR 4895**	44	#2 Lead	260	1878	**29**	Good
375 H&H	IMR 4198	33	#2 Lead	260	1662	31	Acceptable
375 H&H	**SR 4759**	32	#2 Lead	260	1722	**26**	Acceptable
375 H&H	**Re 7**	30	#2 Lead	260	1542	**27**	Good
375 H&H	**Blue Dot**	22	#2 Lead	260	1571	**29**	Good
375 H&H	**Unique**	16	#2 Lead	260	1278	**24**	Good

375 H&H	5744	34	#2 Lead	250	1775	31	Good
375 H&H	5744	40.5	JSP	220	1973	33	Excellent

Caliber	**Powder**	**P Grains**	**Bullet**	**B Grains**	**FPS**	**SD fps**	**Pressure**
308 Win	**SR 4759**	16	RNSP	100	1496	**28**	Good
308 Win	**IMR 4198**	26	FMJ	110	1931	**26**	Good
308 Win	**IMR 4198**	25	JHP	130	1955	**28**	Acceptable
308 Win	**SR 4759**	21	FMJ	150	1586	**25**	Excellent
308 Win	**SR 4759**	22	JSP	165	1615	**23**	Excellent
308 Win	IMR 4198	24	JRN	180	1551	32	Good
308 Win	IMR 4198	24	JSP	200	1482	34	Acceptable
308 Win	Green Dot	11	#2 Lead	110	1699	33	Good
308 Win	Red Dot	10	#2 Lead	110	1587	32	Good
308 Win	**PB**	10.5	#2 Lead	110	1555	**29**	Excellent
308 Win	**Unique**	12	#2 Lead	110	1806	**24**	Excellent
308 Win	700X	9	#2 Lead	110	1521	35	Acceptable
308 Win	**Herco**	14	#2 Lead	110	1863	**28**	Acceptable
308 Win	SR 7625	12	#2 Lead	110	1689	36	Acceptable
308 Win	**Red Dot**	10	#2 Lead	150	1452	**30**	Excellent
308 Win	700X	9	#2 Lead	150	1355	31	Good
308 Win	**Green Dot**	10.5	#2 Lead	150	1400	**28**	Excellent

308 Win	**PB**	10.5	#2 Lead	150	1438	**24**	Excellent
308 Win	IMR 4227	18	#2 Lead	150	1688	37	Acceptable
308 Win	SR 4756	12.5	#2 Lead	150	1496	31	Acceptable
308 Win	SR 7625	11.5	#2 Lead	150	1462	35	Acceptable

Caliber	**Powder**	**P Grains**	**Bullet**	**B Grains**	**FPS**	**SD fps**	**Pressure**
308 Win	Red Dot	10	#2 Lead	170	1293	31	Good
308 Win	Green Dot	11	#2 Lead	170	1389	32	Acceptable
308 Win	**Unique**	10	#2 Lead	170	1320	**24**	Excellent
308 Win	SR 7625	10	#2 Lead	170	1199	37	Acceptable
308 Win	SR 4756	12	#2 Lead	170	1399	34	Acceptable
308 Win	**PB**	9	#2 Lead	170	1148	**30**	Excellent
308 Win	Herco	12	#2 Lead	170	1506	33	Excellent
308 Win	Red Dot	9	#2 Lead	190	1178	35	Good
308 Win	**Green Dot**	10	#2 Lead	190	1209	**25**	Good
308 Win	700X	8	#2 Lead	190	1155	33	Good
308 Win	**PB**	9	#2 Lead	190	1139	**27**	Excellent
308 Win	**Unique**	9.5	#2 Lead	190	1158	**20**	Excellent
308 Win	SR 7625	10	#2 Lead	190	1214	34	Good
308 Win	**Herco**	12	#2 Lead	190	1333	**29**	Good
308 Win	**SR 4756**	12	#2 Lead	190	1255	**27**	Good

Caliber	Powder	P Grains	Bullet	B Grains	FPS	SD fps	Pressure
45-70	IMR 4895	42.4	RN	500	1263	36	40,000 cup rifles
45-70	**IMR 4064**	43.4	RN	500	**1195**	38	40,000 cup rifles
45-70	**SR 4759**	25.5	HP	300	1244	**22**	40,000 cup rifles
45-70	**IMR 4227**	27.2	HP	300	**1179**	**30**	40,000 cup rifles
45-70	RL 7	31.6	HP	300	1248	33	40,000 cup rifles
45-70	**IMR 3031**	40.9	HP	300	**1192**	**19**	40,000 cup rifles
45-70	**IMR 4227**	25	JSP	350	**1168**	**29**	40,000 cup rifles
45-70	**IMR 4198**	30	FSP	400	**1158**	**27**	40,000 cup rifles
45-70	**H 322**	37	FSP	400	**1042**	**25**	40,000 cup rifles
45-70	SR 4759	26	#2 Lead	325	1296	**22**	40,000 cup rifles
45-70	Varget	41	#2 Lead	500	1333	38	40,000 cup rifles
45-70	**IMR 3031**	38	#2 Lead	500	**1188**	**29**	40,000 cup rifles
45-70	IMR 4198	28	#2 Lead	500	1201	35	40,000 cup rifles
45-70	**RL 7**	28	#2 Lead	500	**1143**	**27**	40,000 cup rifles
45-70	**SR 4759**	22	#2 Lead	500	**1088**	**19**	40,000 cup rifles
45-70	**SR 4759**	26	#2 Lead	400	**1139**	**17**	40,000 cup rifles
45-70	**SR 4759**	28	FSP	350	1321	**27**	40,000 cup rifles
45-70	**SR 4759**	28	FSP	400	1349	**21**	40,000 cup rifles
45-70	**SR 4759**	19	#2 Lead	322	**1022**	**14**	Excellent

45-70	IMR 4198	23	#2 Lead	322	1018	18	Excellent
45-70	IMR 3031	29	#2 Lead	322	1035	22	Excellent
45-70	SR 4759	18	#2 Lead	420	933	19	Excellent
45-70	IMR 4198	21.5	#2 Lead	420	961	17	Excellent
45-70	RL 7	33	#2 Lead	420	1173	25	Good
45-70	IMR 3031	34	#2 Lead	420	1140	21	Excellent
45-70	Unique	12	#2 Lead	293	1157	12	Excellent

Caliber	Powder	P Grains	Bullet	B Grains	FPS	SD fps	Pressure
45-70	Unique	9.5	#2 Lead	322	955	19	Excellent
45-70	Unique	13.7	#2 Lead	400	1113	24	Excellent
45-70	Unique	11	#2 Lead	420	987	30	Excellent
45-70	Unique	12.5	#2 Lead	322	1155	24	Excellent
45-70	Unique	10	#2 Lead	385	930	9	Excellent
45-70	Unique	10	#2 Lead	426	843	14	Excellent
45-70	Unique	11	#2 Lead	451	944	19	Excellent
45-70	Unique	10	#2 Lead	500	843	29	Good
45-70	5744	22	#2 Lead	500	1055	22	Excellent
45-70	3100	40.5	#2 Lead	500	991	26	Good
45-70	4350	37.8	#2 Lead	500	1024	29	Good
45-70	8700	54	RNSP	500	899	33	Good

45-70	**H 4198**	29	FNSP	400	**1055**	**24**	Excellent
45-70	**Unique**	11	FSP	400	**747**	**27**	Good
45-70	**IMR 4198**	28	FSP	400	**1032**	**29**	Good
45-70	IMR 3031	34	FSP	400	**1009**	31	Acceptable
45-70	**SR 4759**	26	#2 Lead	292	**928**	**28**	Good
45-70	**IMR 3031**	38	#2 Lead	292	**966**	**24**	Acceptable
45-70	**IMR 4198**	28	#2 Lead	292	**1118**	**15**	Excellent
45-70	**IMR 4198**	21.5	#2 Lead	405	**721**	**20**	Excellent
45-70	**IMR 3031**	34	#2 Lead	405	**1052**	**26**	Good
45-70	**Relo 7**	33	#2 Lead	405	**1333**	**29**	Good

Caliber	Powder	P Grains	Bullet	B Grains	FPS	SD fps	Pressure
45-70	**630**	16.5	#2 Lead	322	<1100	**22**	<10,000
45-70	**SR 4759**	19	#2 Lead	322	<1100	**14**	<10,000
45-70	**IMR 4198**	23	#2 Lead	322	<1100	**11**	<10,000
45-70	**IMR 3031**	29	#2 Lead	322	<1100	**23**	<10,000
45-70	**SR 4759**	19	#2 Lead	366	<1100	**12**	<10,000
45-70	**IMR 4198**	23	#2 Lead	366	<1100	**15**	<10,000
45-70	**IMR 3031**	30	#2 Lead	366	<1100	**26**	<10,000
45-70	**SR 4759**	18	#2 Lead	420	<1100	**17**	<10,000
45-70	**IMR 4198**	21.5	#2 Lead	420	<1100	**17**	<10,000

45-70	**IMR 3031**	34	#2 Lead	420	<1100	**27**	<10,000

Caliber	Powder	P Grains	Bullet	B Grains	FPS	SD fps	Pressure
30-30	**Unique**	7.5	#2 Lead	115	1433	**22**	22,452 CUP
30-30	**Red Dot**	7	#2 Lead	115	1381	**24**	25,600 CUP
30-30	**Green Dot**	7.5	#2 Lead	115	1401	**28**	25,100 CUP
30-30	**Unique**	7	#2 Lead	169	1194	**26**	19,995 CUP
30-30	**Red Dot**	6.5	#2 Lead	169	1133	**18**	24,800 CUP
30-30	**Green Dot**	7	#2 Lead	169	1109	**27**	24,100 CUP
30-30	**SR 7625**	6.5	#2 Lead	169	1086	**21**	22,005 CUP
30-30	**SR 4756**	7	#2 Lead	169	1135	**14**	18,112 CUP
30-30	**Trail Boss**	8	#2 Lead	173	1027	**16**	32,008 CUP
30-30	**Unique**	7	#2 Lead	173	998	**29**	Good
30-30	**SR 4756**	7	#2 Lead	173	982	**26**	Good
30-30	H 335	22	#2 Lead	125	1097	33	17,213
30-30	H 4198	11	#2 Lead	170	1138	38	13,525
30-30	**Varget**	16	#2 Lead	170	1333	**21**	16,661
30-30	**H 414**	23	#2 Lead	170	777	**17**	12,000

Caliber	Powder	P Grains	Bullet	B Grains	FPS	SD fps	Pressure
30-06	**XMP5744**	25	JHP	130	1791	**21**	Excellent
30-06	**XMP5744**	26	FMJ	150	1888	**19**	Excellent

30-06	**SR 4758**	21	JSP	165	1602	**16**	Excellent
30-06	**IMR 4198**	26	JRN	180	1689	**22**	Excellent
30-06	**IMR 4198**	27	JSP	200	1701	**24**	Good
30-06	**Red Dot**	11	#2 Lead	113	1554	**18**	Excellent
30-06	Green Dot	13	#2 Lead	113	1622	33	Good
30-06	Herco	14.5	#2 Lead	113	1779	31	Good
30-06	**PB**	12	#2 Lead	113	1665	**28**	Acceptable
30-06	**Unique**	14	#2 Lead	113	1894	**19**	Excellent
30-06	700X	10	#2 Lead	113	1593	36	Acceptable
30-06	SR 7625	12.5	#2 Lead	113	1656	33	Acceptable
30-06	**Red Dot**	11	#2 Lead	122	1605	**20**	Excellent
30-06	Green Dot	13	#2 Lead	122	1751	31	Acceptable
30-06	Herco	14.5	#2 Lead	122	1818	33	Acceptable
30-06	**PB**	12	#2 Lead	122	1610	**24**	Good
30-06	**Unique**	13	#2 Lead	122	1799	**16**	Excellent
30-06	700X	10	#2 Lead	122	1555	38	Acceptable
30-06	**SR 7625**	12.5	#2 Lead	122	1644	**30**	Acceptable
30-06	2400	18	#2 Lead	122	1662	39	Acceptable
30-06	**RX7**	21	#2 Lead	122	1621	**27**	Good
Caliber	**Powder**	**P Grains**	**Bullet**	**B Grains**	**FPS**	**SD fps**	**Pressure**

30-06	**Red Dot**	11	#2 Lead	151	1605	**22**	Excellent
30-06	**Green Dot**	12	#2 Lead	151	1751	**29**	Excellent
30-06	**Herco**	13.5	#2 Lead	151	1818	**30**	Acceptable
30-06	**PB**	11.5	#2 Lead	151	1610	**24**	Good
30-06	**Unique**	12.5	#2 Lead	151	1799	**18**	Excellent
30-06	700X	10	#2 Lead	151	1555	35	Acceptable
30-06	SR 7625	12.5	#2 Lead	151	1644	31	Acceptable
30-06	2400	21	#2 Lead	151	1662	37	Acceptable
30-06	**RX7**	23.5	#2 Lead	151	1621	**24**	Good
30-06	**SR 4759**	21.5	#2 Lead	151	1689	35	Acceptable
30-06	**SR 4756**	13.5	#2 Lead	151	1556	31	Acceptable
30-06	**IMR 4198**	25	#2 Lead	151	1704	**28**	Good
30-06	**IMR 3031**	31	#2 Lead	151	1749	**10**	Excellent
30-06	8700	54	RNGC	152	1748	**14**	Excellent

(Note: 8700 above is a slow burning, but compressed load)

Caliber	Powder	P Grains	Bullet	B Grains	FPS	SD fps	Pressure
7.62 x 39	Unique	8	#2 Lead	113	1548	36	Acceptable
7.62 x 39	Unique	8	#2 Lead	130	1560	36	Acceptable
243 Win	PB	8	#2 Lead	90	< 1400	33	Acceptable
243 Win	700X	7.5	#2 Lead	90	< 1400	34	Acceptable

243 Win	SR 7625	8.5	#2 Lead	90	< 1400	31	Acceptable
243 Win	PB	8	#2 Lead	95	< 1400	33	Acceptable
243 Win	**700X**	7	#2 Lead	95	< 1400	**24**	Good
243 Win	**SR 7625**	8.5	#2 Lead	95	< 1400	**21**	Good
243 Win	**Red Dot**	8	#2 Lead	95	< 1400	**26**	Good
243 Win	**SR 4756**	10	#2 Lead	95	1498	**27**	Good
223 Rem	**Green Dot**	6	# 2 Lead	54	< 1700	**21**	Good
223 Rem	**Red Dot**	5.4	# 2 Lead	54	< 1700	**22**	Good
223 Rem	**Unique**	6.5	# 2 Lead	54	< 1700	**28**	Good
223 Rem	700X	5.4	# 2 Lead	54	< 1700	34	Acceptable
223 Rem	**PB**	6.1	# 2 Lead	54	< 1700	**21**	Good
223 Rem	**SR 4759**	11.3	# 2 Lead	54	< 1700	**27**	Acceptable
223 Rem	**SR 7625**	6.4	# 2 Lead	54	< 1700	**25**	Good
Ultra Reduced Examples (Tested in T/C)							
308 Win	**Unique**	4.5	SWC	90	989	**18**	Excellent
30-30	**Unique**	4	SWC	90	963	**24**	Acceptable
30-06	**Unique**	5	SWC	90	1000	**22**	Good
308 Win	**Red Dot**	3.3	SWC	90	881	**26**	Acceptable
30-30	**Red Dot**	3	SWC	90	957	**25**	Good
30-06	**Red Dot**	3.5	SWC	90	869	**21**	Excellent

Primer Only Loads (Hammond System-- Brown Nail Gun Primer)

(See pictures in illustration chapter for this system)

30-30	Brown	N/A	Lead Ball	53.4	628	**18**	Excellent
375 H & H	Brown	N/A	Lead Ball	94.4	583	**22**	Excellent
30-06	Brown	N/A	Lead Ball	52.1	711	**19**	Excellent
308 Win	Brown	N/A	Lead Ball	53.9	688	**21**	Excellent
416 Rig	Brown	N/A	Lead Ball	122.3	555	**24**	Excellent
416 Rig	Green	N/A	Lead Ball	122.3	778	**20**	Excellent
375 H & H	Green	N/A	Lead Ball	94.4	691	**19**	Excellent

Note: Nail gun primers: Gray 1, Brown 2, Green 3, Yellow 4
(Not recommended w/o lube/gas checks)

CHAPTER 7. STARTING YOUR BUSINESS

Getting into Reloading or load development as a business requires some advanced planning. With the current anti-gun legislation in many States, and in DC, you will most likely need a manufacturing FFL to sell ammo commercially in large volumes for profit. BUT there are many other ways to make great profit in reloading and ballistics without selling ammo, as hinted at in the niche chapter!

The FFL can also be bypassed under certain federal, military and police contracts if you are organized as a testing lab or exclusive contractor or load developer, as ATF is allowed to make exceptions, and does on many occasions when DOD taps them on the shoulder. Other manufacturers can also contract with you as a lab under their FFL in some cases.

Of course, if your loading is just a hobby, or casual selling to friends, clubs, etc. you CAN'T get an FFL, because it requires a profit motive. In that case, you should still get insured, and get a State Tax (or business license where required), and a Federal EIN (Free Employee Identification Number) -- even if you don't have employees. These allow you to buy wholesale at many sellers, and to open bank accounts, etc. It also is good to go to LegalZoom.com and set up an LLC-- to protect your family and assets against suits.

You can't ship ammo via US Mail-- you'll have to use a common carrier like UPS or Fedex. If you're shipping powder, you'll also need to pay a Hazmat fee to the carrier, and get a signature on the receiving end. This is in the post 9/11 world; if there is another terrorist attack, things will likely get even tighter.

There are a number of insurers listed in Handloader magazines, Shotgun News, etc. Plan on paying as much as $200 a month for insurance, depending on your volume and customer base. You'll also need a good attorney to write the required disclaimers and terms, so someone doesn't come after you for their own misuse of your loads.

For more information, check out the resources chapter below, and don't scrimp on a good accountant and attorney when getting started! **For FFL types and choices, visit www.FFLAnswers.com.** Remember, the rapidly growing field of gallery loading is poised to experience dramatic expansion, and many aspects need new sources, suppliers, research and ideas. If you love guns and reloading, you can be on the leading edge of innovation, whether you're 25 or 65! This includes unique new supplies, testing, formulas, development, tools, equipment and ideas. The Engineering chapter gives a taste for some of the more sophisticated needs out there today. People who work on unique new needs, research, tools, recipes, formulas, ballistics, designs, pressures and applied and basic science are the same people who extended that knowledge to rocketry and the exploration of space, our military's success, the defense of freedom, and safer communities.

Even if you happen to be anti gun, or anti hunting (which begs the question of why you're reading this!), **there is much room in gallery loading for less than lethal loads, stun weapons, animal control, etc. that allows threat management without all the negative legal consequences that come with lethal force**. In addition, home and personal defense tactics of the future, especially for multiple assailants and intruders (which are unfortunately more the rule than the exception today), sometimes call for Military/LE tactics of multiple silent shots that don't give your position, action or strategy away in defending your loved ones. There is a whole field of training based on quiet loads, some classified, and it is important that citizens become aware of the potential benefits of reduced loads in home and personal defense, both in non lethal and quiet yet lethal loads, especially with either unarmed, or multiple armed assailants. This is true in the country, and even more so in the city, where the best defense is often the one that **does not attract undue attention from additional assailants.**

Chapter 8. About Smith Tactical and AAGSR

Smith Tactical was formed in 1971 to serve the research needs of Law Enforcement, Military, Covert Ops, Distance and Precision Shooting, Hunters, Recreational Shooters, and other confidential areas, focusing on the unique niche of **reduced recoil, specialty, and silent yet extremely accurate and uniform load and weapon systems development.**

As in all areas of research and development, the first focus is on safety, and no load development, ballistics or research lab has done as much in advancing the understanding of subsonic and reduced load science variables as Smith. See the Engineering chapter for a taste of some of the more advanced topics, including Smith's proprietary "portfolio" scoring model approach to gallery loads.

After decades of research involving tens of thousands of rounds, and with the aid of state of the art science, instrumentation, math and risk models, and lab facilities, Smith has not only developed a large number of proprietary loads, integrated custom systems, and products, but works on a contracted basis with many of the other top R & D labs worldwide, as well as companies, academic institutions, and organizations.

If you've got a product or project that needs analytic help, investment, development, new energy, evaluation, testing, validation, further research, or just fresh new ideas, email Smith at Mail@SmithTactical.com. We gladly consider RFP's and other contractual solicitations, and are often the low cost provider for even very sophisticated projects, often cutting both time and cost in half with innovative ideas and approaches. Smith also offers inexpensive lab testing and analysis if you don't want to invest in PC, oscilloscope and other technologies at the outset of your business.

AAGSR-- The American Association of Gunsmiths and Reloaders (www.AAGSR.org) is one of the leading associations worldwide, and are innovators in ballistics, load development, and the rights of gunsmiths and reloaders.

Chapter 9. Resources for Fun and Profit

After over 30 years in this business, we've compiled what used to be called a rockin Roll-O-Dex of resources for your benefit, **whether you're loading and researching for fun, or to start or boost a great new business.** We're here to help and support your success!

The "old days" are now known as texts, emails, links, twitter, and web sites, so we're listing the resources below with as much up-to-date info as we can. **Visit www.ReducedLoads.com frequently** for more updates, war stories, and **new resources for the success of your business,** as well as news and reviews of developments in the industry and our electronic newsletter.

We can't restrain ourselves from making some comments, both pro and con, about these resources, to help save you time, money and energy in looking for the right place, for the right resources, at the right price! Please also see Chapter 13 for photo illustrations of some of the resources listed on the websites.

Chapter 10. A Note on Ammo Banners and Regs

In some ways, even though we are living in an age ripe with potential for innovation, we're also living in dark times for shooting enthusiasts. Many in government support the far liberal left that would like to see North America turn into Australia, with NO access to guns, ammo, reloading and cartridge development, research and testing.

The problem is, **most of the innovation COMES FROM private individuals, small firms, and hobbyist experimenters!** If we destroy our young reloading industry, we'll again become hopelessly indebted to foreign powers for that industry.

Reduced loads (RL's) have an important role to play in stemming, then turning this ominous tide. Here are a few reasons why:

1. The majority of **banners hold their views out of ignorance**, and simply following others without ever checking it out for themselves. RL's give many more people an opportunity to enjoy shooting recreationally, becoming comfortable that a gun is simply another tool like any other that has to be treated and operated with skill and respect.

2. Every time there is a "senseless" shooting, the liberal media is all over the tool, and NEVER gives the TRUTH that if the sheep "victims" were armed, there would be NO successful wolves. "GUN FREE" zones (even Fort Hood!) have created more massacres than any other policy! Whenever a CCW law is passed, crime drops significantly. In places like Australia and UK, the public is like sheep or children, completely dependent on a slow responding government to protect them. RL's offer many less than lethal options for those who can't bring themselves to take a life, but also want to protect their families from vicious

predators. Anyone living in reality knows that 911 is really dial-a-prayer, and the police are really there to document what happened to you.

3. RL's are great for youth programs! Our youth represent the future of our art, science, industry and the jobs it creates. If you do only ONE thing with your reduced loads, include a young person in your off times! What will happen to the security of our nation if we "outsource" our defense industries, as liberals would have us do??

4. In a VERY BIG sense, gun issues come down to a battle between the city and the country. City folk don't have to cope with bears, mountain lions and rattlers, but they do have to cope with gun wielding thugs. In our view, a gun is the solution to both-- no one is a victim of either a lion or a thug if they can shoot back accurately. RL's obviously offer very quiet alternatives, as well as less than lethal options, to blasting away in Times Square with a 460 SW Magnum.

WHAT CAN WE DO? Two things:

1. Join NRA.

2. Support and visit www.MOREBANS.org

NRA supports **our rights**, More Bans **tracks State and National efforts to overtly or quietly sneak bans in.** More Bans has a section we all MUST read on ammo-- the terrible State by State effort to **force manufacturers to make only serialized ammo**, which will destroy the reloading industry, as well as huge taxes and fines for **having certain amounts or kinds** of ammo or reloading components.

The real agenda is to create a new "NFA ACT" for ammo, and have "class III" requirements for all ammo components, including primers and smokeless powder. VOTE OUT anyone who supports those initiatives, regardless of party!

CHAPTER 11. ADDITIONAL EDUCATION

Visit www.CertifiedReloader.com,

www.Becomeagunsmith.com,

and www.AAGSR.com

for frequent updates on the best education and certification for gunsmiths, reloaders and a variety of certifications, from new ballistics labs to contracted gallery load tools, equipment and resources.

Chapter 12. Thoughts for Techies, Engineers and Researchers

This chapter is about a few advanced topics in reduced load science and accuracy topics, including safety and detonation models. You can skip it if you hate math and chemistry, or if internal ballistics bore you! **There are many items in this manual for any level of reloader, including beginning researchers**. You might enjoy peeking at some of the more advanced science, research and models here, even if we don't get to every concept in RL's that's being researched today.

To start a reloading, ballistics lab, or advanced gunsmithing business, you MUST be the most informed contributor in your niche! Much of the information in this manual is intended to make you just that-- **THE GO TO person in reduced loads and their accuracy/ safety science**, technology and practical applications. This is true **whether you're selling loads, tools, equipment, or offering your expertise in ballistics, accuracy, personal defense, LE and military, hunting, or load development**.

Explosive Detonation Model

Assume a simplified regression model Y=X1M1+X2M2+X3M3... + R-- (error term). This model has 9 primary dependent variables, which provide a very good predictive fit (low error term) between uniform and non uniform pressure curves (traces). When we refer to pressure curves, traces and waves in this section, we're talking about the *ballistic results of instrument measured pressure events in firearms, not the pressure wave of a detonation, which we cover later.*

Over decades of testing, we've isolated these 9 primary indicators of both excessive pressure, and potential supersonic deflagration (detonation), **using ultra reduced loads**. Almost universally, except for ignoring the weight of the tuft itself in your recipe, **using a small Dacron tuft to concentrate the powder at the primer, and using crimped lead projectiles (or non crimped lead balls), stops this excess pressure process**, and greatly reduces the risk indicators. We also are holding many of the "gun" and advanced physics oriented variables constant in this overview, as many of them have no practical application; email us or check the Reduced Loads website if you'd like more details.

These constants include advanced pressure wave harmonics and dynamics, fluid dynamics of deflagration, thermodynamics of pressure waves, and on the firearm side, barrel whip/ receiver-barrel motion; chamber/ throat/ joint design trade-offs; barrel elasticity, vibration and rigidity, and advanced aspects of ***pressure or combustion wave differences caused by the interaction of bullet concentricity, case neck tension and percent case capacity. (Although our model has a complete list of variables, for those into math, we treat them like partial derivatives, solving each one at a time, while holding the others constant).***

To keep our terms straight, when we talk about subsonic deflagration, we usually refer to pressure or combustion waves, and when we're talking about supersonic detonation, we call them detonation waves (or their generated shock waves). BOTH propagate through the energetic medium and its surrounding environment.

In the regression model, numerous additional variables need to be held constant when evaluating pressure traces for risk, such as double charges, obstructed barrels, eroded barrels, copper and lead fouled barrels, overheated barrels, bad headspace, defective brass, and hundreds of other **causes of excessive pressure that have no or little**

causal or conditional relationship to the reduced nature of the load (other than loader error), which is our isolated topic in this particular model. *Of course the total propellant volume, even of the reduced load, is a potent contributor to the overall deflagration,* ***or more correctly excessive pressure*** *results, but as effect, not cause, so that also is assumed to be a constant. As we'll see below, volume effects blast radius, but not wave speed. Speed is a fixed property of the chemistry of the energetic material.*

Once those variables are isolated, the remaining indicator variables become greatly simplified, and include contributing categories like **powder type, location and speed; projectile type and security; primer selection, bullet type, case neck tension, and relationship of bullet ogive to lands**.

Please remember the **difference between conditional and causal relationships.** Even with years of ANOVA research, most relationships end up conditional rather than definitively causal, because of the number of other variables being held constant, as well as co-variate problems in isolating event horizons that often happen near simultaneously, or at least in minute fractions of a second.

Because of this complexity, in propellant physics and regression studies, even though the variables are all deterministic, they are treated as stochastic due to the impossible number of combinations, rather than variables, that crop up when you begin to hold less variables constant.

Finding out what's "going on" is secondary (as pure research) to the practical *predictive ability of a model who's objective is to avoid risk*. New numerical modeling methods, proton radiography, advances from supercomputing models in shock/ pressure wave harmonics physics, and many other developments in

understanding the physics of propellants vs. explosives, are giving us unprecedented new models for deflagration harmonics undreamt of even in the past 5 years. The recent application of numerical analysis to partial derivatives in explosive/propellant/pyrolant wave dynamics is solving (with supercomputers) problems that were intractable even three years ago!

These new models support our risk algorithm, which theorizes that high pressure high speed deflagration (ie: the "pipe bomb" phenomenon, SEE, or excessive contained pressure) in ultra reduced loads is a *harmonic chain reaction between neck tension/ release, partial bullet motion and blockage via primer, and delayed, non-concentrated (shaped) (and hence oscillating and building) multi-phase powder burn,* when the powder is spread across a case with a relatively large volume of air in contact with the burn surface.

None of our traces have ever supported the idea that reduced loads truly go supersonic or generate actual detonation pressure waves! The traces do support a "shaped" (jet) type wave, similar to a Munroe Effect, which we'll mention below, although remaining subsonic. They DO support the idea of NON UNIFORM pressures, amplified pressures, and secondary spikes which can be "off the scale" (over 100,000 PSI) compared to normal deflagration, but still quite tame (albeit equally deadly) compared to supersonic detonation.

This model supports the fact that Dacron, proper primer choice, crimping, and the use of lead vs. copper jacketed bullets, decreases risk exponentially. The "blocking" and obturation characteristics of copper vs. lead needs further study, as partial bullet motion and blockage, with ensuing high deflagration pressures and pressure spikes, are not as problematic with lead as with copper, for unproven reasons, in addition to anecdotal, intuitive and obvious hardness and willingness to "give" differences.

In that spirit, we have further refined the regression model to create a "portfolio theory" of variables, **where each element contributes both to success and risk.** The problem with anecdotal information in past research (double vs. single base, **both blamed**; slow vs. fast burn, **both blamed**; high bulk vs. fine/close to primer, **both blamed**-- etc.) is that **risk and success were seen as opposites**, when in fact, **greater risk means more and more success**, right up to the point of disaster, just like a financial portfolio! **We NEED uniform HIGH pressure to AVOID the start/stop second amplified pressure wave that is at the heart of detonation**. Yet, too much, and problems multiply and cascade into focused pressure jets and high pressure deflagration (combustion) waves. *This has its correlate in accuracy trade-offs, where much higher pressures (as in magnum cartridges) mean more and more accuracy, right up to the point where* ***barrel vibration makes .223 a much better tack driver than a 416 Rigby!*** *Removing the* ***crucial barrel vibration problems with reduced loads*** *is one of the interesting aspects of its many applications.*

Another problem with past approaches is that the major indicator of uniformity was Standard Deviation (or Extreme Spread in pre-chrony circuit times). **While very important, it has to be augmented with instrumentation, pressure testing, and trace analysis!** Without ballistic instrumentation, pressure spikes can be self cancelling statistically, and mask dangers that SD's ("math") alone will not capture, except perhaps as curious outliers, often discarded as a "thrown" shot. We'll also see, below, how SD's can become curiously "tame" in "shaped" (directional wave or indexed) charges. (We distinguish and compare vector or tensor moment or direction of force in an RL (which we call a pressure jet), from the "shaped" charges of explosives, which we'll cover later).

The loads included in this manual were extensively tested and modeled BOTH for SD AND pressure profiles! Over 160 PUBLISHED, "TESTED" reduced loads were ELIMINATED which had *good SD's, but iffy pressure profiles and dangerous harmonics (traces)! You are reading this right-- not every published reduced load is a safe load, and factors beyond any publisher's control are NOT "held constant" in the real world! (Eg: gun condition, age of powder, uniformity of shell, bullet or hull, external temperature, barrel temperature and fouling, case wall, bullet concentricity and run out, etc.)*

Between the risk indicator variables, and the actual destructive event, is a **real life intermediary that can mean a bad day vs. death-- the firearm itself**. True detonation (the supersonic explosive deflagration transition), which we've never documented, or even extremely focused jet pressure spikes, which we have documented many times, in a Thompson Center, for example, with its relatively **light breech/barrel lock**, compared to a lug secured bolt action rifle, can be very different horses, both in overall destruction, and the **trajectory of the shrapnel debris vs. location of the shooter.**

No manufacturer will tell you that a SAAMI rated action at 60,000 PSI will blow up at just under 100,000, because they are (rightfully) afraid that some idiot will try to load up to that level, when some concentrated jet deflagration (extreme pressure build) events can exceed **LOCAL durability at one spot on the firearm**. (Weakest link in the chain logic). On the other hand, how often do we even think about the **vast differences between SAAMI pressure ratings themselves**, ranging from **11,000 PSI in a .410,**

to 14,000 in the same gun with 45 Long Colt, to 28,000 in the 45-70, 38,000 CUP/42,000 PSI in the 30-30, and nearly 65,000 PSI in the 270 Winchester!

One size does NOT fit all-- you need to check BOTH the **load manuals and the SAAMI pressures** to be sure you're remembering the limits of that particular firearm. Some of the newer load manuals no longer publish pressures for liability reasons, **THANK YOU LYMAN AND LEE**, for continuing to do so. Is it a coincidence that many destructive pressure events have happened with 243 Win, when you consider that many loaders consider it a very "hot loadable" caliber up near 60,000 PSI, compared to say, the 444 Marlin at 40,000?

In a pressure profile, some of the pressure is cartridge related (eg: 7,000 PSI or more), and some "follows" the projectile as the propellant burns. A low risk profile shows a steady bell increase and downslope as the bullet, and pressure energy, exit, with no bumps, big sharp off the scale jumps, or dangerous secondary peaks or jet formations. Irregularities in pressure are often, but not always, reflected in irregularities in SD's (Chrony speeds). **If you only take 5 shots, the 6th could be the one correlating SD and pressure curve!** The first two pressure curves tell a very tight story-- yet we still continue on out to at least 60 shots in our tests (n vs. (n-1) in the SD denominator for a little "population" slack).

A quick, uniform ramp up of pressure, nearly simultaneous between primer and propellant effect, followed by vigorous release of the bullet and land engagement, is the essence of safety, and the whole purpose of skilled chamber design and well constructed loads.

If the powder **burns in stages, burns in one place then another, starts and stops**, or potentially creates supersonic-mimicking harmonics (high pressure secondary jet-like combustion waves) much like an initiator in ANFO, or soldiers marching in step on a bridge, or amplifying harmonics in a fluid dynamics closed model bath; **and the projectile also reflects a start/stop action**, the combination can concentrate high pressure in one part of the chamber, **jam the projectile** momentarily creating "pipe bomb" deflagration containment, or release pressure in stages that create a lodged bullet or building, focused pressure wave.

At best, very inaccurate rounds, at worst, a pipe bomb in your hand. Remember, you don't **need** detonation (a supersonic detonation wave rather than a high pressure focused combustion or deflagration wave) to blow up a pipe bomb, or a gun, with a double base powder. What you DO need is an **initiation sequence (or explosive engine) comprised of exactly the right chain of events**. In explosives, this chain of events are the 5 stages of a detonator. In obstructed propellants, these are the focusing chain of primer/ crimp/ airspace/ dispersed propellant/ jammed or turned projectile/ jet shaped combustion wave. ***Explosives and propellants are really no different-- it's just the process and combination that gets you to different destinations! Otherwise, how could one use a double or single base "propellant" in a pipe bomb?***

Pressure traces show this on the first shot if double peaked or off the scale, two if it is a uniformity overlay issue. SD's take many rounds, by definition! Instrumentation data gets a lot of information all at once; math requires many trials, since it's based on statistical indicators and inferences. **When we see non uniform pressure, even with a single published load, it's time to put that load on the iffy list, even before getting to SD's in an explosion chamber or with a remote trigger.** Do chrony all your loads, but remember it is a lagging indicator!

Let's look at this backwards, and *try to blow up our gun.* **First, use a very slow powder, such as 50 BMG. Put, say, 10 grains in a huge case, like a 30-06 or 416 Rigby, and make sure the case is the normal varying thickness, non concentric low quality we see in many factory cases. Don't use Dacron. Shake the powder so it's spread across the brass lengthwise, with most up near the bullet. Use a small rifle primer. Use a jacketed spire point bullet, set it in lightly, and don't crimp it. Set it .005 or so back of the lands, OR right up against them.**

Will your gun blow up? Probably not, at least for about 100+ tries. (We blew up a T/C with a modified version of this formula we won't publish, in only 18 tries. Supersonic pressure wave detonation? No. Enough obstructed jet-focused pressure to create smaller pieces? Yes). But YES, it will eventually either create a squib, damage the gun, or finally result in a pipe bomb. Do the same in a .410 or T/C, and guess what-- MUCH lower pressures needed to destroy the firearm itself, **making the results even more destructive, and less forgiving.**

That one paragraph pretty well explains the 9 key variables without getting into too much math and predictive weighting. Take each in turn:

1. Single base, very slow powder

2. Under 15 grains, or below 50% of minimum load

3. Jacketed Spire Point bullet

4. No Dacron

5. No crimp

6. Improper land seating

7. Anemic primer

8. Large case, poorly manufactured and formed

9. Powder dispersed, or against the projectile (away from the primer)

(NOTICE that these variables hold a number of OTHER problems constant, like headspace, double charges, known problem powders, gun condition, etc.)

Now, let's successively reduce each risk, by the numbers.

1. Now, instead, we use a double base, very fast, wafer extruded powder (like Unique), **to a published reduced load**. 2400 is also double base, but is tubular, and has had problems in some reduced loads, especially in single shot pistols. **Don't use this as an exclusive variable**, because **one of the BEST reduced powders** (SR 4759) is single base/tubular, "relatively slow" (half way position of 100). One of the powders least sensitive to case position and fill is Titegroup, which is double base/spherical! Titewad works GREAT in reduced (subsonic) 12 gauge loads, Red Dot does NOT (high SD's, bumpy pressure traces). **Trail Boss (extruded flake cut perforated high bulk, single base -no nitro-, with burn speeds in the top 20 up near PB and Green Dot), was made purposely FOR reduced loads**. *As we'll see below, chemical composition of an energetic compound is no indicator of the distinction between detonation and deflagration anyway. One person's propellant is another's explosive. Don't tell anyone, or they will be confiscating our Varget next.*

2. We keep grains to published (reduced) loads, not above or below by even a grain, and we hand weigh each load. Dust off that powder trickler!

3. We use a #2 Alloy Lead bullet or pure lead round ball--IOW, a cast bullet. Jacketed bullets can "vibrate" against lands if improperly spaced, and create a pipe bomb with incorrect obturation and jamming, or at a minimum, be a

nightmare to remove if squibbed. There are many good reduced loads for copper projectiles, some seen here, but for most applications, lead is far more forgiving and safe. Pressure problems with reduced loads were almost 6 times as frequent with jacketed factory bullets as cast lead. Remember, it's the ogive, not the tip, that engages the lands. Spire points can be designed, measured and loaded to tip as a factor in OAL, while forgetting where the ogive is relative to the lands. Also see #6 below.

4. We now use 1/4" Dacron wad holding powder against primer. Many problems of varying and non uniform ignition relate to position of powder within the case. As black powder shooters have known for decades (although excessive case space is, granted, not a problem with black powder!), wads/cards are crucial for smooth, uniform burns, even with copious black powder, let alone reduced smokeless loads. But DON'T FORGET that anything added, including Dacron, has to be factored into the PSI equation. This depends on the application, but wads in a variety of positions can add 3,000 PSI, as shotgun loaders know (and that is a HIGH percent of an 11,000 PSI load limit). Shotgun pressures also are sometimes measured in LUP instead of CUP (Lead instead of Copper crush testing) as well as PSI.

5. We crimp or tightly seat the bullet with exact neck tension. Again, the overall process of ignition, bullet release, "jump", land engagement, and obturation are what create **uniform pressures that are properly directed and uniformly released behind the projectile**. The start/stop, click/bang, slightly "off rhythm" harmonics of reduced load plus interrupted acceleration (misplaced bullet vibration momentum, jump, engagement, obturation and motion) are what tend to mimic "pipe bomb" deflagration to detonation transition harmonics, chemistry and physics (supersonic shock wave chain reaction creation). In other words, in ALL our models, **it is never the powder alone "exploding all at once" as armchair logic has it**, but always the **interaction**

of pressures with projectile, air spaces and case, in building and amplifying a series of high pressure combustion waves or harmonics, and resulting in a pressure jet focused in a non-indexed manner on a partially obstructing projectile. This happens in nanoseconds, so is counterintuitive, but pressure traces don't lie!

6. We seat the bullet within .002 of the lands. Just a thick fingernail for sure, but crucial to proper engagement, obturation, powder harmonics and trajectory. All of loading is trade offs, and although **high pressure and tight land engagement can mean tack driving accuracy, lack of a jump with reduced loads can have just the opposite effect-- non uniform pressures, and keyholing inaccuracy!** See the "tool" pictures in Chapter 13 for suggestions on easily customizing your loads to exact land engagement length, with bullet ogive holding and OAL measurement tools that are very inexpensive. (Don't confuse published .006 advice about headspace with .002 discussions of ogive/land distance, that would make the jump way out of spec).

These careful measurements of OAL vs. lands and crimping are a MUST for jacketed, a SHOULD for lead bullets, and a NICE for round lead balls.

7. We use a magnum, or at least large primer. Armchair speculation has it that a magnum primer "flashes" across the dispersed powder, creating an explosion. There is no "across" with a Dacron wad, and full, uniform ignition lowers risk (of a focused high pressure combustion wave jet, not explosion in the sense of supersonic detonation). Consult your manuals for use of magnum primers, and NEVER use a smaller primer than indicated in the recipes.

8. We match the case to the published load, and make sure it's trimmed, fire formed, and matched to the chamber with

ogive/land, headspace, and OAL, allowing bullet seating the proper distance from lands for each cartridge. Needless to say, **bullet concentricity** is even more crucial in reduced loads for **INTERNAL ballistics safety (land engagement vs. jamming), as it is for stability and accuracy externally with full loads**.

9. Of course, we either tip the barrel up before shooting (cowboy loads), OR BETTER, use Dacron, to keep the powder concentrated near the primer.

Accuracy and Safety are Related in Creating Reduced Loads

Beyond fun, hunting, plinking, etc. are several "next level steps" for extreme accuracy-- often associated with Benchrest (BR), Long Distance, Varmint, 50 BMG, and Sniper shooting, to name a few. In the case of RL's, ***accuracy also is a direct correlate and indicator of safety!***

The elements of extreme accuracy are now well known. Not so well known are how they apply to reduced loads! Holding the gun, environment, and shooter constant, RL accuracy naturally centers around the main projectile and internal ballistic elements: **primer, case, powder, bullet, neck tension, quality of dies**. This of course "holds constant" many important "firearm" variables such as barrel whip/ receiver-barrel motion; chamber/ throat/ joint design trade offs; barrel elasticity, vibration and rigidity, etc; as well as gunsmithing and custom manufacturing "techniques" like barrel bedding, moly coating, cryo treatment, re-rifling, trigger work, bolt polishing and customizing, barrel jobs, bolt lock-up, etc. **Barrel vibration is certainly much less an issue with RLs, for obvious reasons.**

Take powder. It is well known that to optimize accuracy, many loaders, manufacturers, labs, teams, competitors and accuracy shooters use the "87 rule" -- trying for 87% case capacity as a general initial target for powder selection. This can actually vary between 82 and 92 percent with relatively accurate factory match rounds. **5 manufacturers we just tested averaged 84%, with most match loads at 86 to 88.**

The resulting technique is simple-- check your case capacity (maximum case grains via water fill), take 87% of that number, then go to the manuals to find a powder that delivers the velocity you need for that caliber, with close to those grains! *Next, vary between 82 and 92 percent until that combination of gun, powder, projectile, distance, velocity, group size, SD, pressure curve and scope settings are optimized. Manufacturers do this for the "average" production chamber, you can optimize / customize it for EACH gun- cartridge combination. DO use the "grain optimizer" formulas (.3 grain differences etc.) as you do this.*

Of course, **if we're reduced loaders, we look at powder in entirely different ways**-- far less than the ideal case capacity, with selection based on speed of burn, safety, often subsonic velocities, then checking the manuals for that caliber and projectile to begin the process of optimizing SD's, groups and pressure traces.

See the photos in the next chapter for many examples of precision reloading tools and techniques. If we are happy limiting our reduced loads to 50 yards (a frequent near zero or PBR for large centerfire rifles), the differences in powder will be negligible-- **a single hole group is still a single hole group.**

In this regard, case neck tension is both a factor in safety and accuracy, as we'd predict, and of course can't be separated from the effect of the powder on the projectile.

In many ways, all the other elements of attention to accuracy in reduced loads are designed to ***make up for the lower***

powder charge case capacity *in RL's.* ***This is the primary trade off*** *of getting less noise, recoil and expense, at the cost of less than optimal case capacity, while still attempting single hole groups at 50 yards, and MOA at 100.*

Cases-- in RL, it is crucial to use all the calipers, run-out/ concentricity and precision forming techniques we can, **as well as create optimal case neck tension, case uniformity, and consistent dead on headspacing**, if we want superb accuracy as well as safety. The only limit is budget-- there are custom die makers that will provide world class precision if we can afford it, often for hundreds or thousands of dollars each. On the other end of the spectrum, even inexpensive carbide dies, and even normal Lee dies at far lower cost, can provide very tight tolerance brass with the proper technique, maintenance, cleaning, and measurements. The photos next chapter show examples of tools that can check headspace, and design the cartridge length for our exact chamber, as well as assure perfectly uniform brass thickness and concentricity, and neck size rather than full size once the case is fire formed. *Crimping techniques and options alone could be a whole book in itself! An optimal crimp is 10 times as important, where indicated, in RL's as in full starting or max loads, for both safety and accuracy reasons.*

Many anecdotal reports of supposed ultra reduced detonation ended up with shells that weren't available or in good enough shape for forensic examination. How many of these were headspace/ separation issues? If wall thickness was already an issue due to poor case manufacture or bad headspace stretching, how did it contribute to the destructive event? Also, see below for the additive effect case thickness differences can have on the pressure profile itself, let alone catastrophic head separation. Couple that with a firearm with a relatively flimsy breech lock like a T/C, and... you get the idea. (We love T/Cs, but let's stay in reality about the lock design and limits! Lugs anyone? Some of those garage manufactured 50BMG single shots without lugs are a rocket pointed at your chest waiting to go off for the same reason).

Bullets-- Casting ourselves, using round projectiles (a GREAT alternative, see Chapter 13), correct barrel break in, proper crimp tension and placement, properly cooled and cleaned barrels, and checking for concentricity are all key to stable, accurate projectiles. Gauging for run-out in the finished cartridge (both shell and bullet) also is a must. **Over HALF of even the best factory match bullets have concentricity problems today.** A secret many don't know about-- a completely CLEAN and properly BROKEN IN barrel can make the difference between a **good load and an extraordinary load!** We're not talking just Hoppes 9 and lead, we're also talking about something much more subtle-- COPPER. Bullets will not engage rifling properly with copper build up, and "Sweets" type (AK) solvents MUST be used to keep the bore clean of subtle build up, especially during initial break in. This basically means solvents containing ammonias.

If we get a new gun, shoot 15 to 20 rounds, and clean the copper out thoroughly after EACH shot, we'll have **fire conditioned the new manufacturing sharpness out** of the land/ groove transitions enough that copper accumulation shouldn't be an ongoing problem. The more precision the gun, the more important this break in is. **The incredibly built Noreen and Steyer 50 BMG's can easily be 1/2 MOA and better systems-- or 2 MOA and worse, depending on break in! A couple recent articles in rifle mags have challenged this view, but our experience over thousands of tack driving custom sniper and hunting platforms says break in is still crucial.**

Primers-- We exclusively use CCI, and make our own, because many of the others have dropped in quality over the past 10 years. Especially in reduced loads, primers are

primary bullet motion starters, as well as burn initiators, and a poor primer can dramatically affect the most important reduced load safety and accuracy factor--pressure curve uniformity. In our research, **secondary spikes come both from dispersed powder AND poorly functioning primers.** A great primer and that Dacron tuft can combine to give significantly better SD's and PC's than full factory loads, if we are careful with the other factors!

What kits, tools, equipment and other helpful products, information, research and services can YOU come up with, given the above??

When Accuracy, SD and Pressure Deviate: "Shaped" Charges

See the illustration photos and captions in the next chapter, and the resources chapter, for information on tests and systems described in this section. Also see below for more information on the Munroe Effect.

One of the theories that has been in and out of favor over the past 15 years about ultra **reduced load detonation** theorizes that **pressure waves with destructive harmonics can develop when the powder is distributed** across a long line across the shell with a volume of air over it. This is of course not possible with Dacron, but let's explore this for a moment, as there are some exciting and interesting insights to be gained, even though there are flaws in the theory as it is currently stated.

Some call these "shaped charges," which is not in keeping with actual shaped charges (see a bit on explosive chemistry below), and should more properly be called "non-indexed charges" in our opinion. The idea is that uneven pressure is applied due to the string of powder against the opposite wall, which is intuitively picturesque, but completely different than modern numerical analysis models of real supersonic shaped charges.

Instead, think of a few other related examples. First, most commercial shells today have uneven spots in wall thickness. The illustration photos show some examples of how we use the **RCBS Casemaster, NECO Gem gauge, ultrasound and xrays** to identify case irregularities. When BR shooters with unlimited budgets identify these, they sell them to others. When those of us with limited budgets find them, we either correct the run-out (gross) manifestation of the problem with a little tweaking on the NECO CAT, or with pliers/ vice and a soft cloth, OR, when the case is not subtly bent (no run out) but still variant, we **"index" the shell.**

Indexing is a simple process. First, you check thickness and concentricity on your gauge-- when the needle is glued to zero, then "blips" at a certain point, we mark that point with a dry erase marker. After a few more spins, often checking with a second gauge, process or instrument, if the dry erase marks are consistent, we permanently mark the case.

Indexing now simply means that when we chamber 10 rounds, we align the index marks to 12 o'clock, and get a consistent group instead of either a random (clock position driven) cloud, or a vertical string (90 degree off indices). **Notice that when you chrony, speeds can be relatively consistent, but pressures will be quite different!** That means that Standard Deviations are not sensitive to case thickness "position" or index, but pressure is! Here is an example where SD and group size are NOT correlated due to different underlying variables, yet pressure and group size ARE, because **position is causing the pressure to work differentially in the chamber against the projectile**. This even works when the projectile is a round ball, as we'll see. ***Notice that this doesn't require some exotic pressure wave detonation event or "shaped" explosive charge, as much as differential and off center force on the projectile, even when spherical!***

Next, let's eliminate the case as a variable. Take a perfectly made .410 hull (100% uniform thickness), and place the correct formula of powder without the correct wad, fill or overcard, under a cast 218 grain ball or SWC (1/2 ounce)-- for a homemade ".410 slug." (See pics in ch.13 for some of these systems). Without the proper wad or fill, the balls will create a nice vertical string. With the wad or fill, a nice ragged hole! ***SDs and pressure curves show the same relationship as with varying case thickness.*** Obviously, what's going on here is a **differential pressure wave (subsonic) that is hitting with random force in different places on the projectile, or randomly "indexing" itself against the projectile.** This seems to be sensitive EITHER to powder position OR case thickness at this point, right? But **how do we know it is the underlying wave that's the culprit, since we now have TWO variables, albeit isolated?** How can we check the theory *without case variation OR powder distribution variables?*

The third test is elegant and readily available. Brian Hammond up in Canada (see next chapter photos and resources chapter) sells the "game getter" -- a powderless reduced load system that uses .22 nail gun blanks in rifle cartridges (as large as .416 Rigby) as the only propellant. On top of this, since the stud drivers are rim fire, we have a ready made pressure force "index" -- Brian's eccentrically drilled blank holder rims! To hold the other variables constant, we use the same cast ball projectile, select only perfectly concentric cartridges (again easy since Brian incorporates stainless steel sleeves of uniform thickness), **and fire at different positions "on the clock" by rotating ONLY the charge** (.22 blank position).

As expected, **if we "index" the charge and chamber all the blanks in the same position, we get a perfect single hole group at 25 yards.** By <u>rotating the position</u>, we get a variety of results. There are four levels of charge available in stud blanks: Gray, Brown, Green and Yellow. For simplicity, let's say they produce velocities of 400, 600, 800 and 1,000 FPS. Keeping the velocities constant, yet rotating the position of the charge in 90 degree increments gives the familiar .410 "vertical string." Varying BOTH index and charge gives a random cloud-- **a poor group with greatly varying pressures, yet consistent SDs!** Varying only the pressure (with the charge indexed) again gives a nice ragged single hole.

There IS an interaction between deflagration (flame front) "wave" and charge position, yet it is not obvious with SD's alone! Notice that it will chrony similar SDs in certain combinations, indicate "risk" in other combinations, and yet be completely independent of SD in others. The only real way to see what's going on is pressure wave analysis, which is the only indicator, other than grouping, that shows in detail the **"signatures" of position vs charge volume, and how they affect uniformity. Recall again that when we mention pressure waves in this context, we're talking generally subsonic (and potentially very high pressure) pressure traces, NOT supersonic detonation waves!**

The only explanation for these widely different variables (wall thickness, propellant placement, shaped charge offset) having the same results, is the **orientation of the subsonic pressure wave itself** (or pressure "jet" if you ascribe to a shaped pressure charge model). Wave vs. particle-- same idea. In either case, this has to be at the level of directional pressure vectors, otherwise a completely well packed, maximum load uniform charge, in a varying case, would not show the **same pressure trace and grouping outcomes as ultra reduced loads or powderless "shaped" nailgun propellants!**

Notice that these effects can then either be additive or cancelling. Why has it been so difficult to duplicate "detonation" under lab conditions? Probably because

covariates can drive the wave propagation effects (even if they are subsonic pressure waves or directional jets), and covariates also drive the projectile "pipe bomb" start/stop - obturation effects. When you add that **position can either amplify or cancel the wave effect**, it becomes much clearer how complex these variables are. Yet, using round cast lead balls, using Dacron, and choosing carefully made or indexed cases, can obviously reduce or eliminate this category of risk.

We didn't mention neck tension here, one of the most crucial variables we've found in catastrophic pressures. When you have uniform charges, such as Hammond's "brown" nail gun blanks, yet crimp the ball differentially, the chronys give much higher readings depending on crimp, yet DO NOT change the SD/ pressure/ grouping performance at all.

So, it would seem that the "shaped" charge problem (indexed jet is more accurate), if you still want to call it that, is a factor that non concentrically skews the pressure vector, amplifies it, and if there is already a start/ stop pressure problem (the projectile jumping to lands, being eccentric itself, not being crimped and creating a secondary pressure spike, etc.), once in a rare while, all the factors combine detrimentally and produce amplified (still subsonic) deflagration, due in no small part to the behavior of the projectile itself at the lands. We'll look a little at how this initiation engine relates to actual explosive (supersonic) detonation below. Meantime, we can reduce risk by carefully designing and measuring our RL's to fit chamber and lands-- see Ch. 13 for pictures-- in addition to the other steps (cartridge uniformity, bullet concentricity, lead vs. copper, propellant choice, etc.) we've already mentioned in the model itself.

This engine does compare in many interesting ways to the thermodynamics, chemistry and physics of both pipe bombs and truck bombs, which **require very specific PW initiation sequences to work properly.** However, the "shape" of the propellant is far less the issue (from a true shaped charge wave dynamics standpoint) if case thickness in packed charges and .22 blank propellant jets can create similar jets and PT's on a smaller scale (minus, of course, the offending start/ stop jacketed bullet, or a large enough grain weight to cause serious damage). **But these DO work together, and a non indexed wave might itself contribute not only to the start/ stop of the projectile, but also its position in obturating at the lands, to help complete the little pipe bomb. Of course cast lead is much less likely to completely obturate or "jam" (is more forgiving) than copper-- or so goes the current speculation.** ***Powders specifically designed for reduced loads like Trail Boss DO warn about using only cast bullets-- so there is a little fire with this smoke.***

Give It a Little Air?

Also by definition, deflagration is an explosion in which the shock wave arrives before the reaction is complete (because the reaction front moves more slowly than the speed of sound in the medium, and deflagration in propellants is intentionally with incomplete oxygen), producing high heat and moderate pressure. Does attaining supersonic deflagration (detonation propagated with shock compression) in RL's then have its own typology categorized by position, speed, harmonics, powder chemistry, projectile obturation (source of compression?), **air availability or position**, and other factors that drive the timing and position of the waves producing the harmonics? And, even if you never reach true detonation, pressure spikes in excess of 130,000 PSI can still destroy the majority of firearms, and have a "blast radius" of 4 meters, much like a small pipe bomb!

Of course, when we're talking sub vs. supersonic, we mean the speed of the flame front itself. At low speeds, the front produces mostly heat in the heat/pressure mix. Toward the speed of sound, pressure takes over as the primary transition between deflagration and detonation. Energy release (technically enthalpy) comes both from the burn and the conversion from solid to gas. **Some researchers have blamed air volume in the cartridge, both via position and quantity, in this transition. The University of Michigan "deflagration to detonation" study is often invoked for this theory.** ***However, that study was about dust explosions in mines!***

Here's the logic: As you approach the speed of sound, the heat balance with pressure approaches detonation. It never reaches it if pressure stays in balance and is released smoothly via uniform mechanics. However, if there is a sufficient volume of AIR, there can be a heat/air additive pressure event (known as thermal expansion), which when combined with the pressure of the high speed flame front, initiates a supersonic pressure wave (ballistic experts call this a shock wave, explosive chemists and physicists call it a pressure wave, with the shock wave being the secondary wave caused by the PW), which when combined with a certain bullet position, theoretically causes shock compression propagation detonation.

How immediately "available" does this air have to be to the powder, and thus the flame front, in this model? Does our Dacron tuft keep it distant enough from the flame front to keep pressures from building, or is the fact that it is there at all sufficient? If it's a necessary but NOT sufficient condition (excess air space in the cartridge), then Dacron is one part of a simple solution to lowering risk.

Remember, propellant deflagration **IS designed to be an incomplete oxygen event**-- hence the "fireball" at the end of the barrel (known technically as the heat of afterburn) when remaining unburned propellant finally has enough oxygen to finish burning. This actual "SEE" also is seen in cannon fire, sometimes INSIDE the barrel, where there obviously is another large volume of air to produce a secondary deflagration. Notice that a SEE (secondary explosion) as a fireball, vs. contained behind the projectile, is the difference between pretty fireworks and a missing hand.

We also haven't explored the chemistry of RL's vs. different powder types (a bit on explosives chemistry is below, with some surprising conclusions about the irrelevance of chemistry to deflagration vs. detonation)-- and other theories abound that the primer and initial burn start a partial-combustion-products gas fill with certain powder types, which (mixed gas cloud) then detonates like an "air fuel bomb", before the burn is complete. The pressure per grain (max about 1,000 PSI/grain) vs. potential psi in the detonation is ignored in this model, assuming that the SPEED of the reaction itself can produce in excess of 100,000 PSI, even with grains as low as 3 (breaking the laws of thermodynamics for most modern smokeless powders-- anyone loading with pure HMX or RDX?), supposedly before the bullet can even move (as this questionable theory goes), let alone jam/ tilt. **Remember, the extent of damage in any detonation, (or more correctly in this case, a non indexed, jet pressure spike) is primarily a function of 1. The total amount of fuel burned in the event (total energy available); 2. The maximum flame velocity that is achieved; and 3. The manner in which the expansion of the combustion gases is contained.**

(Important Note: Just because we've found that few to no excessive pressure events in RL's are true (or at least complete) detonations, doesn't mean that they aren't extremely dangerous. In fact, a high pressure obturation that turns your gun into a small "pipe bomb" can be **MORE dangerous than a LARGE DETONATED pipe bomb!** If you pack a 3" closed steel pipe with Nitrocellulose, Potassium Nitrate and Sulfur, then initiate it with a blasting cap, it has a kill radius of over 18 feet. A much larger pipe bomb, though very destructive, tends to blow upwards, with less anti personnel effect. *Gun-turned-pipe-bomb is non-trivial, even when it is not technically a detonation).*

If there was enough "air" (key to air fuel bombs!) to cause such a phenomenon, cowboy action powders like Trail Boss

(as low as 3.5 grain recipes pushing a 245 g cast bullet) would be blowing up all over the land, since the powder has incredibly low bulk density (high fill) at .328, compared to 1+ g/cc in modern slower burning rifle powders! This is so "fluffy" that IMR can get only 9 ounces in Hodgdon's normal 1 pound container! Seem like plenty of "air" (even if dispersed between the tiny donuts?) to cause detonation trouble? IMR doesn't think so! We're talking about a powder **designed for** sub 800 FPS loads here. *Someone was very smart at IMR and figured that a lot of the "SEE" events were likely double or triple charges.*

This theory would also suggest that air OUTSIDE of densely packed high density powder (front of shell in dense starting loads), or air DISPERSED between low density donuts of powder (Trail Boss) are both fine, but **air partially but not completely mixed** can be a problem. We generally doubt that some air fuel event is an issue at all, given the above. After all, the Nitro in a double base powder IS one source of the oxidation!

Also, don't forget that the slowest burning powders, which often **take more volume than the fast powders, have been *implicated far more often in unusual pressure spikes*** than the fast burners taking far less space, or leaving much more "air" room-- going against the "grain" of the airspace argument. **The extra "room" in the cartridge is much more related to the non indexed jet formation** in our research (see the **Munroe Effect** below), than the extra oxygen/nitrogen (air) in the case, otherwise, Dacron wouldn't prevent the problem (which it does, as a propellant concentrator and indexer, NOT an air displacer). Our "jet" theory is not a unique idea, although it is a result of our research and that of our Explosive Technologies sister company-- as a subset of the too general "shaped charge" idea-- see the Munroe Effect discussion below.

Putting one of Hammond's Yellow nail blanks in a 30-30 case would be the ultimate "air" problem-- and it's not, as long as you "index" the pressure jet consistently from shot to shot, depending on caliber! In the case of the 30-30, there is no offset by design, so no manual indexing needed, but still, all the rest is air, and Yellow stud drivers can deliver supersonic projectile speeds. Accuracy is surely much better with Brown and Gray stud driver velocities, but with round lead projectiles, no issue with excess pressure has ever developed in thousands of rounds, for nearly 20 years.

Both Dacron and lead projectiles certainly seem to point to the true source of destructive pressure events-- a non indexed jet that causes or is combined with a stuck or partially start/stopped projectile, most often jacketed. If lead, consider that many other factors can cause obturation, including heat, a dirty or worn barrel, etc. If powder is indexed and primer oriented with Dacron, consider other variables that can destroy indexing, like non concentric and non uniform cases and bullets, as well as bullet and neck run out.

Notes on Explosive Chemistry vs. Propellants

We all know from High School chemistry that work comes from conversion of stored to kinetic energy, by releasing the energy in ionic or covalent chemical bonds. We won't get into updates on kinetic frames of reference, invariant mass, etc. that relegates KE to the observer watching the bullet pass. All combustible energetic materials (CEMs) have stored energy, and all are capable of combustion, deflagration and detonation with the right combination of initiating and propagating circumstances.

A deflagration, as in a firearm chamber, is a combustion wave propagating sub-sonically with high pressure, whereas a detonation creates a detonation wave propagating at supersonic speed. All CEMs have oxidizer and fuel components (eg. propellants and explosives), and hybrids (pyrolants-- some classified as primaries, pyrotechnic or primer explosives), which often also have tiny metal particles added. **Explosives must burn under detonation with high enough velocity to create a shock wave. Rocket propellants must deflagrate fast enough to generate high exhaust velocity through constricted nozzles. Firearm propellants must deflagrate fast**

enough to create high pressure, without creating a supersonic detonation wave.

In older research, propellants were in fact called "low explosives." They were categorized this way initially because of having low oxygen rather than low speed (enough to deflagrate but not detonate), and were looked at as "surface burners." The only way a propellant could detonate under the early formulations and models was as a secondary consequence of **gaseous product build up under confinement** (strictly a pressure phenomenon). Double bases and other "high" oxygen (and speed!) propellant formulations were found to be able to **detonate with shock initiation by a primary explosive,** just like a "high explosive," *blurring if not eliminating the distinction under confinement conditions.* Technically, **without projectile blockage (or start/stop projectile "slowing" obturation), the confinement requirement for detonation is not met** in either the gaseous pressure (jet), or the initiation models.

Explosives are categorized as primary and secondary--secondary (like TNT, ANFO, RDX, HMX, etc.) requiring strong initiation to explode, and primary (and in some cases Pyrolants) (metal or ammonium nitro phenols, diazodinitrophenol, or mercury fulminate etc. --used in primers and initiators), which detonate easily with heat, fire, impact, sparks, etc. (It's well known that you can burn TNT with a blow torch-- an explosive train initiation is needed for secondary explosives, as opposed to a simple fuse, impact, heat, spark, etc. for the primaries). Some primaries/ pyrolants actually create no gas, unlike propellants and secondary explosives.

An **explosive train initiator** (such as a blasting cap), which initiates a stable, powerful secondary explosive, generally has 4 or 5 sections. At the top, a crimped rubber plug holds a black powder fuse, an electric match, or electrical wires leading to a Ni Chrome bridge wire in the priming mixture (or a chip, in sophisticated fuses). The priming mixture (PM), usually lead styphnate, or mercury fulminate/nitro phenols, is ignited by one of these external inputs. The PM, in turn, ignites a booster charge (usually lead azide), which detonates a secondary and then primary "base charge" -- often made of the same secondary explosive as the main charge, such as HMX, RDX or TNT.

Note that Nitrocellulose (gun cotton, or NC) comes in both explosive and propellant grades, with explosive grades containing a few percent more Nitrogen (in the 10-14% range for both). Of course the application and design (explosive vs. propellant) determines the mixture-- often with Nitroglycerine in double base propellant powders, and with ammonium nitrate, sodium nitrate and saltpeter for blasting powder explosive use. **Like any CEM, NC can be either a propellant or an explosive.** Retardants and surface area/ shape/ size of granule design, as well as lack of permanent containment and quantity, attempt to limit NC CEMs in firearms to their role as propellants.

Technically, if a propellant burns normally, it goes through burn zones known as the Fizz zone, Dark zone and Flame zone, creating "plateau, mesa" and other stages of the burn. Retardants, as well as granule size and shape, as well as CEM recipe proportions, "customize" the burn rate and flame temperature in each zone. **The vast range of burn rates exist to accommodate different bullet weights and barrel lengths**-- it takes faster or slower burn rates to drive heavier or lighter projectiles through shorter or longer barrels of differing calibers. In addition, pressures have to be managed to differentiate between light psi tolerances, as in shotguns, and stronger pressure tolerances, as in magnum rifles.

Next, **say you detonate two charges, one containing 500 pounds of TNT, and one containing 50**. Will the smaller quantity effect the speed of the pressure (detonation) wave? Intuitively, yes, actually, no. It will of course effect the blast radius, but the **SPEED of the pressure wave is always constant for each type of explosive, just like each propellant form has a constant burn rate.** Velocities (m/s), for comparison, of several compounds are: Mercury Fulminate: 3920; Nitrocellulose: 7300; RDX (ie Semtex, C4, Comp B): 8700; TNT: 6900; PETN (technically an unusually sensitive secondary, but used in detonation cords, fuses, primer cords and rocket propellants): 8380; ANFO (truck bombs, demolition explosives): 4100; TATB (used as a nuclear trigger to initiate plutonium fission): 7600.

We can take one of the most commonly used explosives (Ammonium Nitrate/ AN -- or ANFO when mixed with fuel oil, aluminum, etc.), and turn it into a propellant. This is called "AN Black Powder," and is made by mixing in kerosene, then extruding, with sulfur, charcoal, sodium nitrate, calcium carbonate and urea.

To get to the important correlate with RL's-- notice that AN is a very "weak" explosive (even though it brought down a building in the OK bombing when mixed with fuel oil)-- meaning it has a relatively weak pressure wave speed, as compared to many other compounds, including Nitrocellulose. Relatively weaker PW's "propel" debris at high speeds more than "blowing them apart" like TNT and other high explosives. Hence, AN is used extensively in mining, where it is important that rocks are cleared out, rather than blown into bits, causing cave ins. Combine this with the fact that anti personnel pipe bombs work better in smaller (3") form than larger (foot long), due to the shrapnel effect (larger pipe bombs sending shrapnel up more than out, and often in smaller pieces), and we can start to see that, congruent with explosive theory, complete **"demolition detonation" events are not really even needed to create serious injury with a propellant pressure spike.**

Assume we don't have all the explosive train elements in our chamber to create a true, or sustained, detonation wave followed by a shock wave. Assume also that we have a very small charge, which, while not capable of creating a large blast radius, still has the characteristics of high speed PW's (Nitrocellulose and Nitroglycerine in a double base powder). Now, assume we also don't have complete obturation (a closed container), but rather a start/stop blockage that spikes the pressure in stages, just as our SD's and traces show so often in the "questionable" RL's.

The likely result will be a lower velocity PW that tends to rupture the firearm and send the components (rather than small pieces of shrapnel) flying away from the rupture point. Potentially deadly, yes, but far from being the actual equivalent of a truly sealed small pipe bomb containing a similar amount of powder. In making the actual pipe bomb, we've also got to add (we won't give the percents) potassium nitrate and sulfur to the NC. Because the shooter is "holding" the bomb, and due to the small size of the chamber, even though the shrapnel isn't grenade size or velocity, it still may well be in the exact direction of the shooter's face, eyes and hands, rather than straight up or forward, also depending on the design of the chamber and lock, and condition and age of the firearm.

The Munroe Effect

Although DDT and DTD (Deflagration to Detonation Transition and Detonation Transition to Deflagration) have been studied extensively in dust explosions, another much more closely related area to firearm propellants is rocket propulsion.

Some shooters are surprised to find that many of the same components (including Nitrocellulose and Black Powder) have been used in rockets as in firearms for many years. This makes intuitive sense, as we're "propelling" a tiny rocket (our bullet) down the barrel, rather than creating an explosion. Of course volumes are much higher, and pressures much lower, in rockets. Still, explosions in rocket launches also are not intended events.

"Rocket Science" (Thermochemical aspects of deflagration propellants) gives us many insights into both effectiveness and safety in Reduced Load science. For example, a number of studies were done years ago, attempting to correlate smokeless powder's chemical and physical characteristics to the likelihood of SEE or detonation and deflagration pressure spike jet events. As we hinted at above, there is no correlation.

There are two reasons for this: 1. ANY CEM (Combustible Energetic Material) can be EITHER a propellant or an explosive (including the newer category of pyrotechnics called pyrolants)... and 2. It has been proven over many years, in many studies, that although the quantum chemical, physico-chemical and physical chemistry processes of **deflagration and detonation are significantly different**, the two ***cannot be distinguished or separated solely on the basis of the physico-chemical properties*** of the CEM! (Propellant Handbook, JES, Tokyo, 2005 et al, Kubota p. 273, many others). "2" obviously also follows logically from "1" above.

This doesn't mean we can use fast or slow burning powders with disregard or reckless abandon in RL's, but it clearly shows that **other factors in the combustion chain and recipe** MUST be responsible for the uniformity of outcomes, good or bad, *along with* the powder chemistry. It also clearly explains while radically different formulations can work equally well or poorly depending on other factors. **As mentioned earlier, double base, fast burning (top 10), spherical *Titegroup*, and *SR 4759*-- single base, tubular, relatively slow (not in top 40 in speed), are both excellent RL powders.** *This also explains why both fast and slow powders have been called both better and worse in anecdotal advice on forums, and both have been blamed for excessive pressure, many unfairly.*

Our "jet" combustion wave theory of differential "hits" on the bullet base or surface due to shape of the pressure (combustion) wave has a corollary in detonation waves. Called the "Munroe Effect," when an actual detonation is initiated in a charge, a uniform, spherical wave propagates evenly throughout the charge, and if not inhibited, creates a related shock wave in the outside environment. If, however, SPACE exists (like the 42 degree space in front of an inverse V shaped charge), opposite the point of detonation, a high energy, directed jet stream is created, because the detonation wave itself is deformed by the space within the container.

In detonation physics, this is used to cut through thick steel plating. In propellant physics, the pressure wave itself (when seen in our traces as a secondary sharp peak, or fitting to other risk shape indicators-- see below), can, with the aid of the air space in a larger cartridge, and **as an effect rather than a cause of the secondary spike** (we'd assign more causality, which is unknown, to the partial jamming or stopping of the projectile), FOCUS high pressure on one part of the wall, or weakened head, or inadequate locking of the cartridge or action, or differentially on the projectile.

Advanced Research

In both ST and ETI ballistics and explosives labs, **new research projects are applying rocket propulsion numerical modeling to firearm ballistics**. This is only recently possible due to the completion of algorithms for Nitrocellulose within several models, including the BKW (ISPBKW) software programs (eg. Mader, Taylor and Francis, 2008, FORTRAN code available on CD ROM, see resources chapter). Previously, non-ideal gas behavior in NC propellants (though tiny) created uncertainties that ISPBKW and other algorithms remove.

With the new code, which generates Po, CT, ET and Isp (1.808, 3914, 2293 and 296 for NC, respectively), we are now able to create a "piston model" much the same as with rocket propellants, to measure under driven, steady state and over driven pressure waves. *Of course this is only an analog to detonation waves overtaking their own shock wave in explosives-- in both rockets and firearms we (hopefully) only deal with pressure/combustion waves, not detonation!*

Combined with **brand new harmonics algorithms in several recent ballistics and pressure tracing programs** (see resources chapter), concurrent numerical modeling breakthroughs are giving us a much greater understanding of the relative contributions of primer, gas, heat, powder characteristics, case fill, neck tension, projectile type and other variables in the drive states and stages of combustion pressure waves within firearms, **particularly with reduced loads or unique projectiles.**

The immediate practical application of these studies is our ***proprietary rankings, seen in this manual, of previously "acceptable" RL's into more refined categories than SD's alone will reveal.*** Our ultimate goal is to publish risk model AUC's that we and other investigators can match to their own traces, with the objective of including automated risk level analysis in the PT algorithms, and then the recipes themselves.

In the interim, loaders can check their own chronys, SDs and recipes against the indicators we've published here and on the ReducedLoads.com website. ***Powder manufacturers are welcome to contact us if some of their favorite loads are missing, to find out if we haven't yet tested their recipe, or if it indicated waveform hazards, and we quarantined it. We're also happy to test newer combinations for manufacturers, to keep our readers up to date on the performance of newer powders.***

CHAPTER 13. ILLUSTRATIONS-- PHOTOS OF A "BASEMENT BALLISTICS LAB" COURTESY OF SMITH TACTICAL-AAGSR

Special thanks to Smith Tactical for use of the following illustrations/ photos. **Visit www.ReducedLoads.com frequently for updates**, additional load recipes, and the latest research studies and results, as well as success stories and ideas from fellow gallery loaders, and **THANK YOU for your interest in this exciting and growing niche** within reloading, ballistics, research and gunsmithing.

Except for the ballistics microscope and pressure testing equipment, many of the items in the following photos are standard in the labs and shops of advanced load developers. The "hunting pistol" photo shows a variety of large bore handguns from 500 and 460 Smith to 375 H&H, .416 Rigby and a 45-70 "BFR." Not shown (set up in the field) are a number of chronys and bullet traps, as well as stop action photography equipment and electronic microscopes.

Remington
ARSENAL
PRIMERS

poly-fil
Low-Loft Batting
Bonus Stuffing
Puff-Lon
Mix #47 Buffer

(50 BMG)
WILSON
Tools and Gages
FOR ACCURATE —
RCBS
Lyman

DOUBLE CAVITY
LEE BULLET MOLD
DIAMETER .380 DOUBLE
WEIGHT 82 gr.
380 RB
PURE LEAD SHOULD BE
USED FOR ROUND BALLS
7 34307 90423 4
DOUBLE CAVITY
LEE BULLET MOLD
DIAMETER .410
EIGHT 210 gr.
410-210-SWC
LE CAVITY MOLD

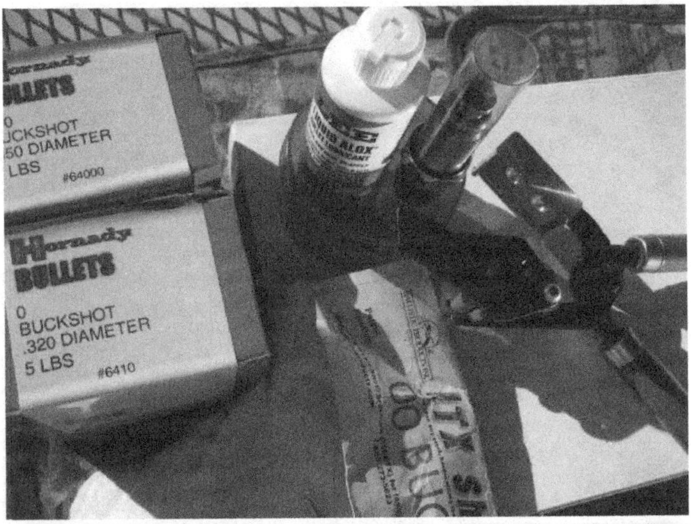
BULLETS
0
BUCKSHOT
.320 DIAMETER
5 LBS
#6410

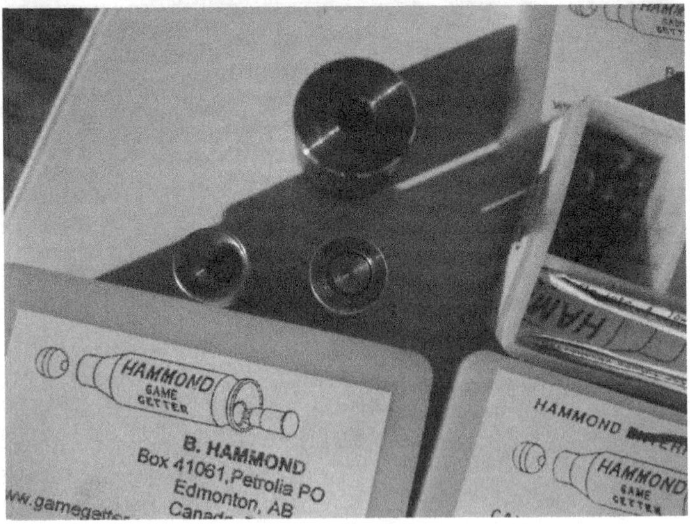
HAMMOND
GAME
GETTER
B. HAMMOND
Box 41061, Petrolia PO
Edmonton, AB

Ramset
.22
100
Ramset
.22
100
416 RIGBY

ALLIANT POWDER
NEW
410
SMOKELESS POWDER
FOR .410 SHOTSHELL
Accurate
4064

IMR
IMR
4895
SMOKELESS
POWDER
DANGER!
HODGDON
WAD
LOADING DATA CENTER

HODGDON
HODGDON
HODGDON
LIL'GUN
RELOADING DATA CENTER
DANGER
IMR
SR
4759
SMOKELESS POWDER

Accurate
2495
Double Base Smokeless Propellant
1680
ALLIANT POWDER
RELODER
22
SMOKELESS MAGNUM RIFLE POWDER
RELODER
15
SMOKELESS MEDIUM RIFLE POWDER

Accurate
495
IMR
SR
4759
.410 Light Game Slug
BERGER

Roll Crimper
12 gauge

Lock-N-Load
BULLET COMPARATOR
& COMPLETE
INSERT SET

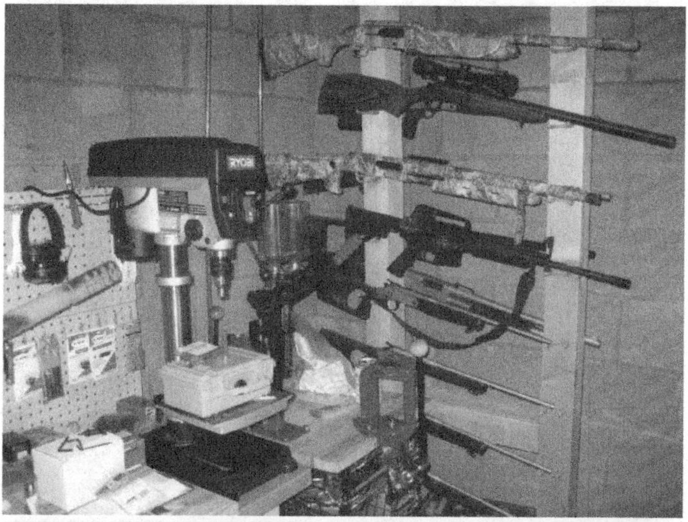
RYOBI

RYOBI

Swift
UNIVERSAL
THREE JAW CHUCK

Remington

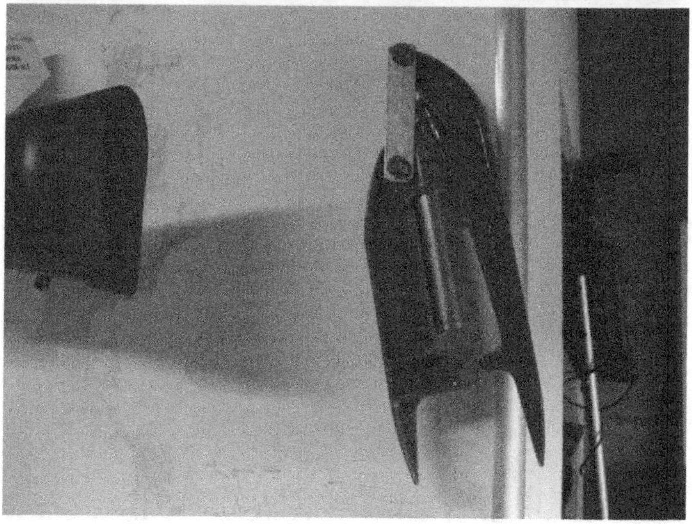

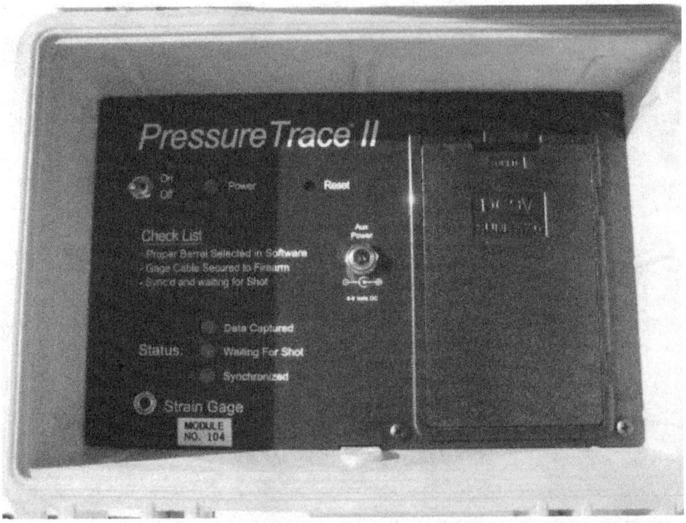
PressureTrace II
On
Off
Power
Reset
Check List
- Gage Cable Secured to Firearm
- Sync'd and waiting for Shot
Aux Power
Data Captured
Status:
Waiting For Shot
Synchronized
Strain Gage
MODULE
NO. 104
DC 9V

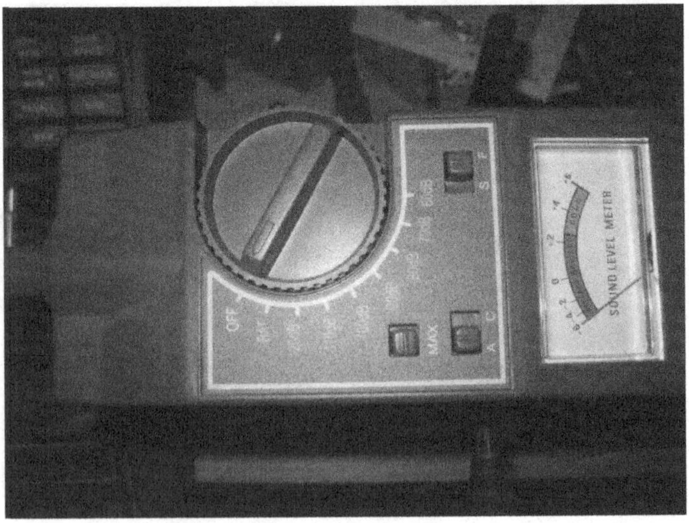
SOUND LEVEL METER
OFF
MAX
A C
S F

RCBS

Chapter 14. Conclusion: A bright future for ballistics research

The research in this report took over 5 years to complete-- for a relatively small population of loads. When strict scientific rigor is applied to longitudinal ANOVA analysis, far more trials are required to reach consistent significance.

There is an astonishing "sidebar" that has not been reported in this study. Along with the gallery loads, your author, Smith Tactical and AAGSR conducted a wide range of "baseline" normative studies, using off the shelf ammunition from well known manufacturers.

While we are critical of variance profiles with SD's above 30 fps, we are continually astonished at the variability of factory loads! SD's of 100, 200 and even 300 are not uncommon! Saying that a correlation between SD and safety is "unproven" is the understatement of the year in this context-- otherwise factory loads would be wildly erratic and causing far more accidents than actually happen.

Even more astonishing is the paucity of good ballistics research. Surveying the literature shows hundreds of great new articles on ballistics, filled with tensor calculus and fascinating deep math. Looking a little deeper, however, shows amazingly few high-trial ANOVA regression models attempting to prove all the correlations we take for granted with all that math!

On one hand, the scientific method of checking every alley just to be sure it is blind is expensive. On the other, the cost of really good testing equipment has come down so much, that it is now within the reach of the individual ballistic researcher, and home labs are multiplying like rabbits. Spending many thousands on EC's, pressure testing equipment, good chronys, ballistic scopes, oscilloscopes, ballistic microscopes, statistical packages, etc. used to be prohibitive. Today, most of these technologies are available in very inexpensive forms that hook to a PC via USB!

This is old news to a staunch group of endangered loaders-- those who have long known that factory loads are inadequate for really accurate results. They don't have to be told that factory cases vary in thickness and uniformity to such a small extent that it is unnoticeable to even MOA target shooters, but to such a large extent that no one betting their life on them should make that bet.

After all the safety disclaimers in this report, not a single detonation event has been conclusively documented that can definitively be causally related to a light load rather than a double load, improper crimp, or other miscalculated overload or pressure spiking variable. That's because "after the fact" the evidence is a mess, and the variables combined.

When we TRY to duplicate these events in controlled trials, after many hundreds of trials, the "controls" themselves seem to prevent the occurrence. Whether this is some mysterious Schrodinger's cat working at a yet unknown Quantum observer level at room temperature, or more likely elimination of the real culprit-- double charges-- is unknown. We all know the car never "makes that noise" when the mechanic is present.

The lack of research is an opportunity for those who wish to pursue a field that is still in its scientific infancy. Our own research in ballistic acoustics has resulted in some astonishing discoveries, such as the Ventriload™, which disguises the location of a sniper team with sound displacement. Here we are well into the 21st century-- where are all the phasers and photon torpedoes? Has the politically correct, anti-gun movement taken the initiative away from our labs, and shifted the momentum to a young Chinese child who tomorrow will invent the next planet shaping technology? Not yet!

Perhaps the most intriguing new research is in projectiles themselves-- smart loads that use a phone app and embedded chip to guide a shotgun missile, for example. New flash-bang, less than lethal and other discoveries are all out there. As in astronomy, it is not always the big budget, government funded lab that makes the breakthrough. As long as our entrepreneurial spirit remains intact, the next major patents will likely come from you, the dedicated research entrepreneur.

About the Author

Mr. Halstead is co-founder of Smith Tactical ballistic research and development labs, and a founding member of AAGSR. His primary field is creation of algorithms and data structures for chipsets and software, including pressure trace and ballistic logic gates, vectors and tensors, although his algorithms have been used extensively in many other fields. These range from medicine and diagnostic devices to virtual lie detection, video game collision detection, crime scene analysis and photography, sound signature location and numerous supercomputing technologies and applications, including RNA folding. His current research interests include Neurology, Artificial Experience systems, training, emotion under stress, and ballistic acoustics. He can be reached via email at Contact@ReducedLoads.com.

Made in the USA
Monee, IL
04 November 2026

67380514R00105